Multifunctional Antimicrobial Peptides

Discovery, Diversity, Biological Activities, Modes of Action and Applications

Aqeel Ahmad and Abdulrahman M. Alshahrani

Copyright © 2017 Aqeel Ahmad and Abdulrahman M. Alshahrani

All rights reserved.

ISBN-13:978-1542336475

ISBN-10:1542336473

Abbreviations

AMPs	Antimicrobial peptides
ATP	Adenosine Triphosphate
ATR-FTIR	Attenuated total reflectance Fourier transform infrared spectroscopy
Bac	Bactenecin
BMAP	Bovine myeloid antimicrobial peptide
CAP	Cationic antimicrobial peptide
CD	Circular dichroism
CD-AMPs	Cathelicidin-derived antimicrobial peptides
CRAMP	Cathelin-related antimicrobial peptide
DNA	Deoxyribonucleic acid
DPC	Dodecylphosphocholine
FDA	Food and Drug Administration
DS	Dermaseptin
MGD1	Mytilus galloprovincialis defensin 1
MIC	Minimum inhibitory concentration
NAD+	Nicotinamide adenine dinucleotide (oxidized form)
NADH	Nicotinamide adenine dinucleotide (reduced form)
NMR	Nuclear magnetic resonance
OaDode	*Ovis aires* dodecapeptide
OaBac	*Ovis aires* bactenecin
PMAP	Porcine myeloid antimicrobial peptide
PR-AMPs	Proline rich antimicrobial peptides
PR-39	Proline-arginine-rich 39-amino acid peptide
PG	Protegrin
RTD-1	Rhesus macaque θ-defensin
ROS	Reactive oxygen species
SDS	Sodium dodecyl sulfate
SMAP	Sheep myeloid antimicrobial peptide
Trp	Tryptophan
TR-AMPs	Tryptophan-rich antimicrobial peptides

CONTENTS

	Summary	v
1	**Introduction**	**1-5**
2	**Discovery and Structural Diversity of Antimicrobial Peptides**	**6-13**

    2.1   Discovery of Antimicrobial Peptides

    2.2   Classification of Antimicrobial Peptides

        2.2.1   α-helical antimicrobial peptides

        2.2.2   β–sheet Antimicrobial Peptides

        2.2.3   extended antimicrobial peptides

        2.2.4   β–hairpin or loops Antimicrobial Peptides

3	**Overview on cathelicidin-derived, Proline-rich and Tryptophan-rich Antimicrobial Peptides**	**14-27**

    3.1   Cathelicidin- derived Antimicrobial Peptides

    3.2   Proline rich Antimicrobioal peptides

    3.3   Tryptophan-rich Antimicrobial Peptides

4	**Biological Activities of Antimicrobial Peptides**	**28-49**

    4.1   Antibacterial Activity

    4.2   Anticancer Activity

    4.3   Antifungal Activity

    4.4   Antiviral Activity

    4.5   Antiparasitic Activity

4.6 Insecticidal and Spermicidal Activities

4.7 Immunomodulation

5 Modes of Action of Antimicrobial Peptides   50-58

5.1 Carpet Mechanism

5.2 Barrel-stave Mechanism

5.3 Toroidal Pore or Wormhole Mechanism

5.4 Aggregate Model

5.5 Shai-Matsuzaki-Huang (SMH) Model

5.6 Self-Promoted Uptake of Cationic Peptides

5.7 Alternative Mechanism of Action

6 Application of Antimicrobial Peptides   59-65

6.1 Therapeutic uses of Antimicrobial Peptides

6.2 Application of Antimicrobial Peptides in Agriculture

6.3 Application of Antimicrobial Peptides in Food Industry

6.4 Antimicrobial Peptide Surface Coating

6.5 Application of Antimicrobial Peptide in Biosensors and Detection

6.6 Antimicrobial Peptides in Drug Delivery System

References   **66-89**

Summary

Increasing conventional antibiotic resistance has created an urgent need to develop novel therapeutic agents. Antimicrobial peptides can be alternative to overcome this problem because of broad-spectrum activities and their unique mode of actions, which minimizes the chance of bacterial resistance. In this book, we have discussed in detail about the discovery, classification, structural and functional diversities, modes of action and application of antimicrobial peptides. Moreover, we have also discussed on cathelicidin-derived, tryptophan-rich and proline-rich antimicrobial peptides about their sources of isolation, structure, function and their applications.

CHAPTER-1
INTRODUCTION

1. INTRODUCTION

Several bacterial pathogenic organisms are present in the environment which is very harmful to our body. Therefore, we need strong immune system in our body, which could have the ability to fight against bacterial infections. Our body produces several antimicrobial peptides (AMPs), which play a primary role in the defense system of the body, for example, our body produces LL-37 AMP, which defend our urinary tract from bacterial infections (Zasloff 2006). Therefore, the demand of antimicrobial peptides tremendously increases in recent years (Hilpert et al. 2005; Oh et al. 2005; Otvos et al. 2005). It is to be noted that not only human beings, but other mammals and several plants and animals produce many AMPs, which protect them from bacterial infections (Broekaert et al. 1997; Hancock 1999; Hancock and Sahl 2006). Interestingly, these peptides are found on those parts of the body which are in direct contact with the environment like nose, ear, skin, eyes, trachea, lungs, and gut (Hancock and Scott 2000; Ahmad et al. 2012).

Naturally occurring or synthetic AMPs are multifunctional (Fig. 1.1) because these peptides doesn't only kill different bacteria but also kill fungi, parasites, protozoa and even the anticancer activity of these peptides have been reported (Lai and Gallo 2009; Scott and Hancock 2000; Otero-González et al. 2010; Yeung et al. 2011; Ahmad et al. 2012). Moreover, the multifunctional role of AMPs makes them attractive candidates for possible use as bio-preservatives, as agents in wound healing, or to enhance disease resistance in aquaculture (Jenssen et al. 2006; Park et al. 2009). These peptides also have several immunomodulatory functions (Fig. 1.1) (Bowdish et al. 2005). In addition, these peptides also used as, antiviral, insecticidal and spermicidal agents (Fig. 1.1) (Reddy et al. 2004a; Zairi et al. 2005; Jenssen et al. 2006; Craik 2012).

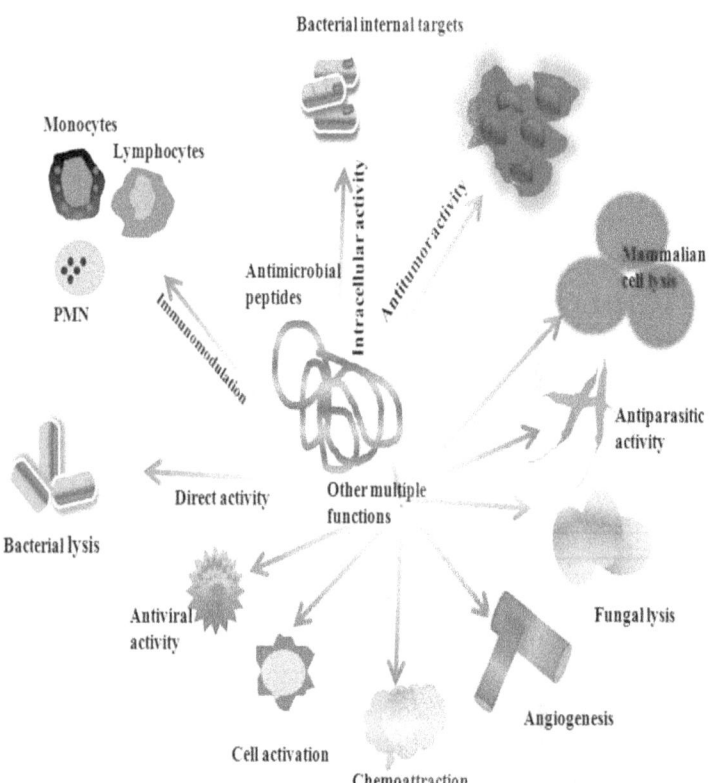

Fig.1.1: Multifunctional role of antimicrobial peptides. See text for details. Adopted from (Ahmad et al. 2012).

AMPs share some common features such as, they are amphipathic and cationic in nature (Nizet and Gallo 2003). Several factors, including size, sequences, charge, structure and hydrophobicity affects their biological activities (Brogden 2005). The structures of several AMPs are analyzed by various techniques including nuclear magnetic resonance

CHAPTER-1: INTRODUCTION

(NMR), circular dichroism (CD), attenuated total reflectance Fourier transform infrared spectroscopy (ATR-FTIR) and X-Ray crystallography (Sitaram and Nagaraj 1999). The literature studies suggested that these peptides formed a particular structure in a membrane mimetic environment or in the presence of lipid vesicles and thus play a key role in predicting the biological activities based on structure. Magainins AMPs (Matsuzaki 1998), isolated from frog, are unordered in aqueous environment but show a helical structure in the presence of membrane mimetic environment or lipid vesicles and also cecropins AMPs (Hwang and Vogel 1998; Soares and Mello 2004) show random-coil structures in aqueous solution but, show an amphipathic α-helix upon in the presence of lipid vesicles (Fig. 1.2). The formation of a particular conformation state is very essential for determining the biological activity of these peptides (Ahmad et al. 2006; Ahmad et al. 2009b; Ahmad et al. 2011; Pandey et al. 2010) . Several peptides formed helical structure in the negatively charged membranes which are well correlated with antibacterial activities and some peptides formed or maintain their helical structure in zwitterionic membrane which is good matches with their toxic activities (Ahmad et al. 2006; Ahmad et al. 2009a; Ahmad et al. 2011; Ahmad et al. 2012) . Despite ample studies, however it is not well understood, how the structural features of these peptides control the biological activity and this warrants further research in this area.

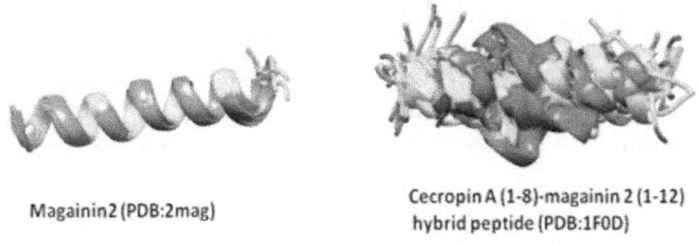

Magainin2 (PDB:2mag)

CecropinA (1-8)-magainin 2 (1-12) hybrid peptide (PDB:1FOD)

Fig. 1.2. Alpha-helical structures of magainin2 (PDB:2mag) and cecropin A (1-8)-magainin 2 (1-12) hybrid peptide (PDB:1FOD) in the presence of dodecylphosphocholine(DPC) micelles.

In recent years, demand of AMP has increased because of the

development of bacterial resistance against the conventional antibiotic (Yeaman and Yount 2003; Brogden 2005; Ahmad et al. 2012). However, the chances of the development of bacterial resistance against AMPs are less because they kill bacteria by physical disruption of cell membrane

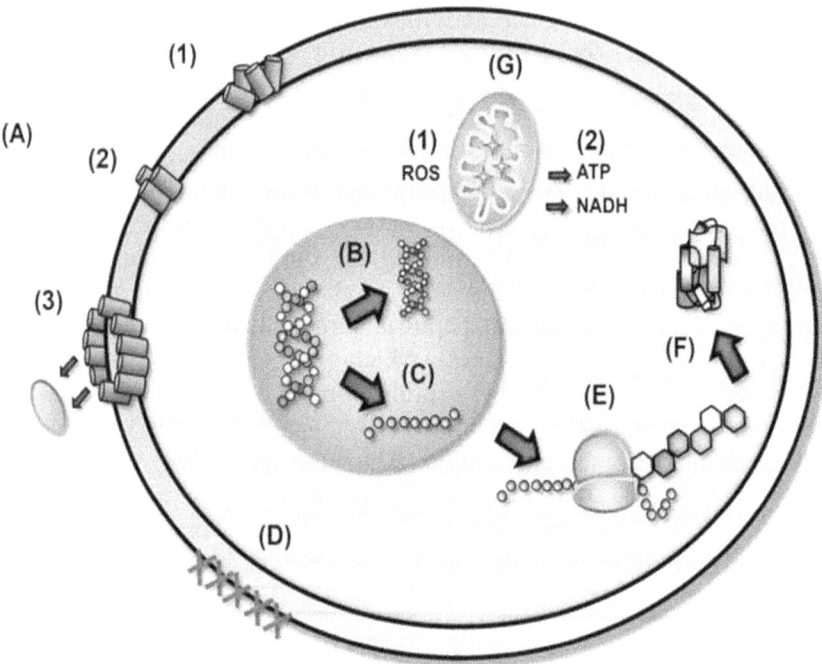

Fig. 1.3. A cartoon showing the modes of action of AMPs. (A) Rupturing of cell membrane (B) DNA synthesis inhibition (C) Enzyme inhibition (E) Inhibition of protein synthesis and ribosomal function (F) Inhibiting protein folding by blocking chaperone proteins (G) Killing cells by targeting mitochondrial cells through inhibition of cellular respiration and induction of ROS formation (1) and also by disruption of mitochondrial cell membrane integrity and efflux of ATP and NADH (2) (Adopted from (Peters et al. 2010).

and disrupt the whole integrity of the bacterial cell and thus, leave very little scope for the development of bacterial resistance (Hancock and Chapple 1999; Hoskin and Ramamoorthy 2008) . The AMPs mostly target the membrane of the microorganisms in order to kill them. Many scientists from different parts of world have proposed several models to elucidate the mode of action of different AMPs such as carpet, barrel-

CHAPTER-1:INTRODUCTION

stave and toroidal or worm hole mechanism (Yeaman and Yount 2003; Brogden 2005; Jenssen et al. 2006). Some peptides kill bacteria through "Aggregate model" mechanism or through "Self-promoted uptake of cationic peptides"(Jenssen et al. 2006). We know that conventional antibiotics are very specific in their action. Some antimicrobial peptides also kill the bacteria in a similar way as conventional antibiotic (Hale and Hancock 2007; Nicolas 2009; Ahmad et al. 2012).

Toxicity is first and the major obstacle to convert these AMPs into drugs. The majority of these peptides beyond a certain concentration cannot discriminate between mammalian and bacterial cells. Despite ample studies, very little is known to control the cytotoxic activity of these naturally occurring or designed peptides. Despite the multifunctional role of AMPs, little research has been carried out to evaluate the cytotoxicity of the peptides in mammalian systems. Therefore, development of novel ways for designing of novel AMPs with reduced toxicity is urgently required. Furthermore, to understand the molecular basis of action of these designed peptides or naturally-occurring peptides by studying the interaction of the peptides with model membrane and different bacteria and mammalian cells and also by studying the assembly of the peptides therein is also very important.

CHAPTER-2
DISCOVERY AND STRUCTURAL DIVERSITY OF ANTIMICROBIAL PEPTIDES

2. DISCOVERY AND STRUCTURAL DIVERSITY OF ANTIMICROBIAL PEPTIDES

Several AMPs have been isolated from distinct species and also many peptides have been designed and characterized by using several biophysical experiments. Also structures of these peptides have been resolved by using different techniques. In this section, we will discuss in brief about the discovery of several types of AMPs from different organism and also their classification based on their structure.

2.1 Discovery of Antimicrobial Peptides

AMPs are discovered in nearly all the living organisms including plants, bacteria and animals (Wang et al. 2009). Numerous AMPs are isolated from vertebrate species particularly from fishes, amphibian, birds and mammals and also from invertebrate species which include moth, beetle, spider, fruit fly, mollusk, sea squirt, and crustacean (Bulet et al. 2004). Tachyplesin I, polyphemusin I, II and big defensin AMPs are found in horseshoe crab (Hancock et al. 2006). Mytilin A, and defensin MGD1 are found in mollusk. Sarcotoxin IC, sapecin B, and cecropin have been discovered in flesh fly (Hancock et al. 2006). Heliomycin and defensin are isolated from moth (Hancock et al. 2006). Moreover, gomesin an AMP isolated from spider in 2002, which is active below 10µm concentration against a large number of bacterial and fungal strains (Mandard et al. 2002). Pardaxin P-1, misgurin, pleurocidin, parasin 1, moronecidin and hepcidin are some selected examples of AMPs discovered in fishes and sphe-2 AMP are isolated from birds (Ganz and Lehrer 1998). Some of the antimicrobial peptides isolated from different organism are represented in table-2.1.1.

CHAPTER-2: DISCOVERY AND STRUCTURAL DIVERSITY OF ANTIMICROBIAL PEPTIDES

Table-2.1.1: List of some antimicrobial peptides isolated from different organisms.

Antimicrobial Peptide	Sequence	Source	Reference
LL-37	LLGDFFRKSKEKIGKEFKRIVQRIKDFLRNLVPRTES	Human	(van der Does et al. 2012)
Bactenecin	RLCRIVVIRVCR	Bovine	(Wu and Hancock 1999)
α–Defensins (RK-1)	MPCSCKKYCDPWEVIDGSCGLFNSKYICCREK	Rabbit	(McManus et al. 2000)
Theta-defensin (RTD-1)	GFCRCLCRRGVCRCICTR	Monkey	(Selsted and Ouellette 2005)
Sphe-2	SFGLCRLRRGFCARGRCRFPSIPIGRCSRFVQCCRRVW	Birds	(Thouzeau et al. 2003)
Pardaxin P-1	GFFALIPKIISSPLFKTLLSAVGSALSSSGEQE	Fish	(Thompson et al. 1986)
Magainin1	GIGKFLHSAGKFGKAFVGEIMKS	Frog	(Zasloff 1987)
Gomesin	ECRRLCYKQRCVTYCRGR	Spider	(Mandard et al. 2002)

CHAPTER-2: DISCOVERY AND STRUCTURAL DIVERSITY OF ANTIMICROBIAL PEPTIDES

Plectasin	GFGCNGPWDEDDMQCHNHC KSIKGYKGGYCAKGGFVCKCY	Fungi	(Mygind et al. 2005)
Pyrularia thionin	KSCCRNTWARNCYNVCRLPGTI SREICAKKCDCKIISGTTCPSDYP K	Plant	(Vernon et al. 1985)
Microcin J25	VGIGTPIFSYGGGAGHVPEYF	Bacteria	(Salomon and Farías 1992)
Polyphemusin I	RRWCFRVCYRGFCYRKCR	Horseshoe crab	(Miyata et al. 1989)

Some peptides were also discovered in fungi such as plectasin, (Fig.2.1.1) produced by the saprophytic ascomycete *Pseudoplectania nigrella,* which show promising antibacterial activity against various gram-negative and gram-positive bacteria (Mygind et al. 2005). Further study revealed that plectasin kills bacteria by directly binding the bacterial cell-wall precursor Lipid II (Schneider et al. 2010). Plants also produce many AMPs which protect them from bacterial infection (Marcos et al. 2008). These peptides are found in almost all the parts of plants and expressed constitutively or in response to microbial infection. Moreover, they are small cationic peptides and stabilized by formation of disulfide bridges. Thionins, defensins, cyclotides, glycine-rich proteins, snakins, 2S albumins, and hevein-type proteins are some selected examples of AMPs discovered in plants, which are appreciably active against phytopathogens (Barbosa Pelegrini et al. 2011).

A group of small AMPs called bacteriocins, found in bacteria, are highly effective against microorganisms. These peptides differ in size, structure and mode of action. Bacteriocins usually cause the bacterial cell death through pore formation or by blocking of cell wall synthesis. Bacteriocins isolated from Gram-negative bacteria are known as microcins. Microcins are divided into two subgroups; class I and class II microcins (Duquesne et al. 2007). Class I microcins are smaller in size

CHAPTER-2: DISCOVERY AND STRUCTURAL DIVERSITY OF ANTIMICROBIAL PEPTIDES

and have extensive post-translational modifications, for example, microcin C7 and microcin J25 antimicrobial peptides (Hassan et al. 2012). In other hand, class II

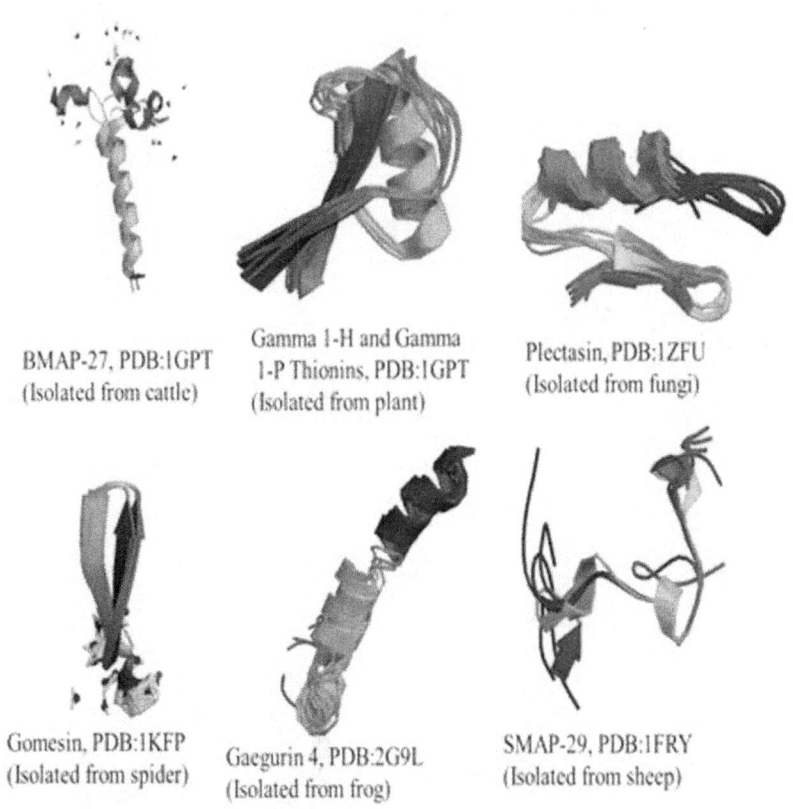

Fig.2.1.1: Representation of structures of some antimicrobial peptides isolated from different organism.

microcins are relatively large in size and have little or no post-

CHAPTER-2: DISCOVERY AND STRUCTURAL DIVERSITY OF ANTIMICROBIAL PEPTIDES

translational modifications, for example, microcin E492, colicin V and H47 (Hassan et al. 2012). Bacteriocins isolated from Gram-positive bacteria are divided into two major classes: lanthionine containing (class I or lantibiotics) and non-lanthionine containing (nonlantibiotics or class II) bacteriocins (Cotter et al. 2005a, b). The examples of lantibiotics includes nisin, subtilin, Mersacidin and Duramycin C. Nonlantibiotics

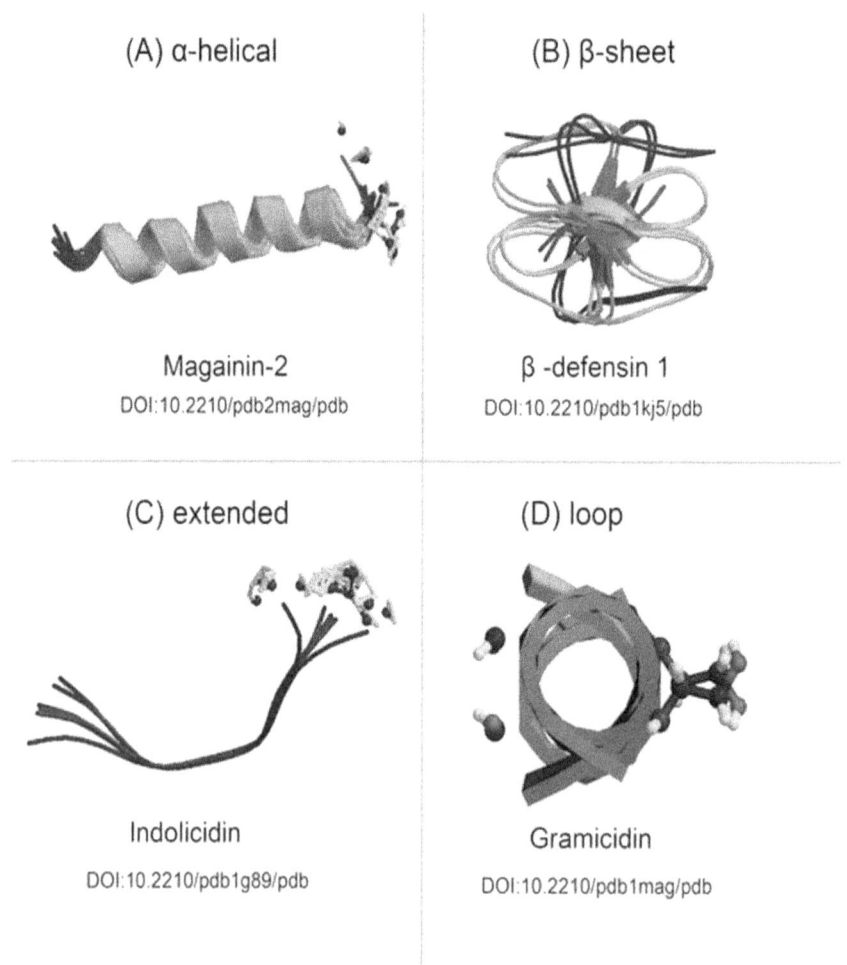

Fig. 2.2.1: Demonstration of structures of AMPs representing the four

CHAPTER-2: DISCOVERY AND STRUCTURAL DIVERSITY OF ANTIMICROBIAL PEPTIDES

classes of antimicrobial peptides (Adopted from (Peters et al. 2010).

are cationic in nature, smaller in size, and heat-stable bacteriocins. For example pediocin PA-1, enterocin A, mesentericin Y105, and leucocin A, enterocin AS-48, lactocyclicin Q, garvicin ML, subtilosin A, sublancin and glycocin F are nonlantibiotics AMPs (Cotter et al. 2005a, b; Hassan et al. 2012).

Several potent AMPs have been identified from different amphibian species. Amphibians skin glands are a rich source of several potent AMPs, for example, magainin and PGLa are isolated from *Xenopus laevis* (Simmaco et al. 1998). Brevinin, tigerinins, and esculentin AMPs are produced by Indian species frog *Rana tigrina* (Sai et al. 2001; Kreil 1994). Some peptides have also been discovered in amphibia of the *Phyllomedusa* genus such as dermaseptin, adenoregulin, and gaegurin (Simmaco et al. 1998). Some frog of Australian species belonging to the genus *Litoria* or *Limnodynastes* produces AMPs including caerin, frenatin, and maculatin (Simmaco et al. 1998). Moreover, buforin I and buforin II are discovered in Asian toad (Rinaldi 2002; Jenssen et al. 2006).

AMPs are found in several species of mammals including human beings, cattles, rabbits, mouse, and monkeys. Mammalian AMPs are divided into cathelicidin and defensin. Defensin is found both in plants and animals, which are classified into α, β and θ defensins (Ahmad et al. 2012). LL-37, hCAT-18, and histatin are found in human beings. LL-37 is one of the extensively studied human AMPs which protects our body by eliminating microbes through diverse mechanisms (van der Does et al. 2012). The study also suggested that LL-37 protects mice against lethal endotoxemia and also prevents macrophage activation by bacterial components (Scott et al. 2002). BMAP-27, BMAP-28, indolicidin, and bactenecin are some well-known examples of AMPs which are isolated from cattle (Ahmad et al. 2009a; Ahmad et al. 2009b; Ahmad et al. 2012). Mammalian AMPs like protegrin-1 from pigs, φ defensin from monkeys, and α–defensins from rabbits, have been isolated in the last few years (Bulet et al. 2004).

2.2 Classification of Antimicrobial Peptides

CHAPTER-2: DISCOVERY AND STRUCTURAL DIVERSITY OF ANTIMICROBIAL PEPTIDES

AMPs differ significantly in their length, amino acid composition, charge, and structures, but despite these huge differences there are many common characteristics among these peptides, for instances, most peptides are amphipathic and cationic in nature and also all of them exhibits antibacterial activity. Some AMPs are unstructured in aqueous solution, but form significant secondary structures in the existence of a membrane mimetic environment or in the existence of lipid vesicles. Accumulating data suggested that the formation of a particular structure is required for the biological activity of antimicrobial peptides. It is important to comprehend the structure-function relationship to design relevant peptides for various therapeutic applications.

The above mention text already indicates that the AMPs form a different type of structure depend on the environment in which they exist. Based on their secondary structures, AMPs are divided into following four major classes: α-helical AMPs, β-sheet AMPs, extended AMPs, and β-hairpin or loops (Fig.2.2.1) AMPs (Powers and Hancock 2003; Mcphee and Hancock 2005; Ahmad et al. 2012).

2.2.1   α-helical Antimicrobial Peptides

It is one of the major class of AMPs. The members of this class form α-helical structure. Several peptides belong to alpha helical AMP class, for example, cecropins, which is isolated from moths and flies and melittins, found in bee venom, which exhibits broad-spectrum antimicrobial activity, but are also highly toxic towards mammalian cells (Pandey et al. 2010). Some toxins such as lycotoxins isolated from the venom of the wolf spider *Lycosa carolinensis* and pardaxins found in the sole species, are alpha helical AMPs (Yan and Adams 1998; Bhunia et al. 2010). Pleurocidin, misgurin, and parasin are alpha helical antimicrobial peptides, isolated from fish (Tossi et al. 2000). Some alpha helical peptides are also found in amphibians such as magainins, esculentins, gaegurins, rugosins, caerins, maculatins, frenatins and dermaseptins (Tossi et al. 2000). Also mammals produce a number of AMPs which belongs to this class, for example, SMAP-29, isolated from sheep, show a broad spectrum of activity against bacteria, fungi, and viruses, also form significant alpha helical structure (Fig.2.2.1) (Dawson and Liu 2009).

2.2.2   β–sheet Antimicrobial Peptides

CHAPTER-2: DISCOVERY AND STRUCTURAL DIVERSITY OF ANTIMICROBIAL PEPTIDES

It is also one of major class of AMPs, which are characterized by presence of β-sheet secondary structure (Reddy et al. 2004b). Some peptide like cecropin found in silk moth, magainin found in the frog, and bactenecin 1 from cow show beta sheet structure (Mcphee and Hancock 2005). Other AMPs like thanatin and lactoferricin B also adopt β-sheet structure (Mcphee and Hancock 2005). Defensins are small and cationic peptides found in several organisms including human beings which are sub divided into α-defensins, β-defensins and θ-defensins, also belong to β-sheet AMPs group (Fig.2.2.1) (Pazgier et al. 2007; De Smet and Contreras 2005).

2.2.3 Extended Antimicrobial Peptides
This class of peptide lacks secondary structure and often contain high proportion of certain amino acids such as arginine, histidine, proline and tryptophan (Ahmad et al. 2012). Some members of this class do not show membrane activity and cross the biological membrane like cell penetrating peptides and target cell components intracellularly. Proline rich AMPs like pyrrhocoricin, drosocin, and apidaecum penetrate the cell membrane and interact with intracellular proteins (Nguyen et al. 2011). Histatins also belong to this class, isolated from human salivary glands which contain a very high proportion of histidine residues (Oppenheim et al. 1988). Indolicidin, a tryptophan rich isolated from the cytoplasmic granules of bovine neutrophilsm, is also a good example of extended AMP family, which is shown in the fig. 2.2.1 (Peters et al. 2010).

2.2.4 β–hairpin Or Loops Antimicrobial Peptides
The peptides of this class contain one or more disulphide bond and form relatively rigid structures (Ahmad et al. 2012). These peptides show high stability because of the existence of disulphide bonds between the β-strands. Tachyplesins, polyphemusins and bactenecin are some examples of this class of peptides (Laederach et al. 2002; Powers et al. 2004; Mcphee and Hancock 2005; Ahmad et al. 2012). Gramicidin AMPs also forms loop like structure as shown in the fig. 2.2.1 (Peters et al. 2010).

CHAPTER-3
OVERVIEW ON CATHELICIDIN-DERIVED, PROLINE- RICH AND TRYPTOPHAN-RICH ANTIMICROBIAL PEPTIDES

3. OVERVIEW ON CATHELICIDIN-DERIVED, PROLINE RICH AND TRYPTOPHAN-RICH ANTIMICROBIAL PEPTIDES

AMPs are very diverse in nature and therefore it is not easy to categorize them into a very specific group. Despite these diversities, they have some common characteristic. For Example, Cathelicidin-derived antimicrobial peptides (CD-AMPs) are characterized by the presence of cathelin domain. Some AMPs are classified into groups because of the existence of certain amino acids in high proportion. For instance, Proline-rich antimicrobial peptides (PR-AMPs) contain a high proportion of proline residue and tryptophan-rich antimicrobial peptides (TR-AMPs) contain a high proportion of tryptophan (Trp) amino acid in the sequence. In this section, we will discuss in detail about some important group of antimicrobial peptides such as CD-AMPs, TR-AMPs and PR-AMPs.

3.1 Cathelicidin- derived Antimicrobial Peptides

Cathelicidin are a family of endogenous antimicrobial peptides which exert diverse functions and also show a broad spectrum of biological activity (Nizet and Gallo 2003). They have been isolated from several species including human beings. The first antimicrobial peptide was isolated from mammals, but later on it was discovered in many organisms such as fishes, birds, amphibians and reptiles (Ahmad et al. 2012). In mammals, they are isolated from humans, monkeys, mice, rats, rabbits, pigs, cattle, sheep, goats, and horses (Gennaro and Zanetti 2000; Ramanathan et al. 2002). Some of these cathelicidin derived antimicrobial peptides are represented in table-3.1.1. They are isolated

CHAPTER-3: OVERVIEW ON CATHELICIDIN-DERIVED, PROLINE- RICH AND TRYPTOPHAN-RICH ANTIMICROBIAL PEPTIDES

Table-3.1.1: Selected examples of cathelicidin-derived antimicrobial peptides

Antimicrobial Peptide	Sequence	Source	Reference
LL-37	LLGDFFRKSKEKIGKEFKRIVQRIKDFLRNLVPRTES	Human	(van der Does et al. 2012)
BMAP-27	GRFKRFRKKFKKLFKKLSPVIPLLHLG	Cattle	(Ahmad et al. 2009b)
BMAP-28	GRFKRFRKKFKKLFKKLSPVIPLLHLG	Cattle	(Ahmad et al. 2009a)
Protegrin 1	RGGRLCYCRRRFCVCVGR	Pig	(Cole and Waring 2002)
Bactenecin	RLCRIVVIRVCR	Cattle	(Romeo et al. 1988)
mCRAMP	GLLRKGGEKIGEKLKKIGQKIKNFFQKLVPQPEQ	Mouse	(Gallo et al. 1997)
PMAP-23	RIIDLLWRVRRPQKPKFVTVWVR	Pig	(Tossi et al. 1995)
PR-39	RRRPRPPYLPRPRPPPFFPPRLPPRIPPGFPPRFPPRFP	Pig	(Gallo et al. 2002; Zanetti et al. 1995)
rCRAMP	GLVRKGGEKFGEKLRKIGQKIKEFFQKLALEIEQ	Rat	(Travis et al. 2000)

CHAPTER-3: OVERVIEW ON CATHELICIDIN-DERIVED, PROLINE- RICH AND TRYPTOPHAN-RICH ANTIMICROBIAL PEPTIDES

Chicken CATH-1	RVKRVWPLVIRTVIAGYNLYR AIKKK	Bird	(Xiao et al. 2006)
Prophenin-1	AFPPPNVPGPRFPPPNFPGPR FPPPNFPGPRFPPPNFPGPRF PPPNFPGPPFPPPIFPGPWFP PPPPFRPPPFGPPRFP	Pig	(Harwig et al. 1995)
HFIAP-1	GFFKKAWRKVKHAGRRVLDT AKGVGRHYVNNWLNRYR	Fish	(Ganz and Lehrer 1998)
Cathelicidin-PY	RKCNFLCKLKEKLRTVITSHIDK VLRPQG	Frog	(Wei et al. 2013)

from several parts of the body such as neutrophils, phagocytes, leukocytes, mucosal cells and skins. They form different type of structures (Fig.3.1.1) and based on these structures, they are classified into following four groups (Risso 2000; Gennaro and Zanetti 2000).

I α–helical cathelicidins
Cathelicidin-derived antimicrobial peptides which form a helical structure belong to this group. Several α-helical cathelicidins have been isolated from humans, monkeys, mice, rabbits, guinea pigs, sheep, cattle, pigs, and horses. Example of this group is LL-37/hCAP-18, BMAP-27, BMAP-28, BMAP-34, OaMAP-34, SMAP-29, PMAP-23, PMAP-36, PMAP, CAP 18, and CRAMP (Zanetti 2004; Zanetti et al. 1995; Boman 2003; Ahmad et al. 2009a; Ahmad et al. 2009b).

II β–sheet cathelicidins
Members of this class form β-sheet structures. Some selected examples of this class include Protegrins.

III Extended cathelicidins
The peptides of this class contain high content of certain amino acids such as proline, arginine, tryptophan and phenylalanine. Some Extended cathelicidins peptides are isolated from cattle (Bac4, 5, and 7), sheep

CHAPTER-3: OVERVIEW ON CATHELICIDIN-DERIVED, PROLINE- RICH AND TRYPTOPHAN-RICH ANTIMICROBIAL PEPTIDES

(OaBac5, 6, and 11), goats (ChBac5) and pigs (PR-39, Prophenin-1).

IV Loop-structured cathelicidins

Cathelicidin-derived peptides which form a loop like structures come under this category. For example dodecapeptide, isolated from cattle and OaDode isolated from sheep myeloid cells belong to this class.

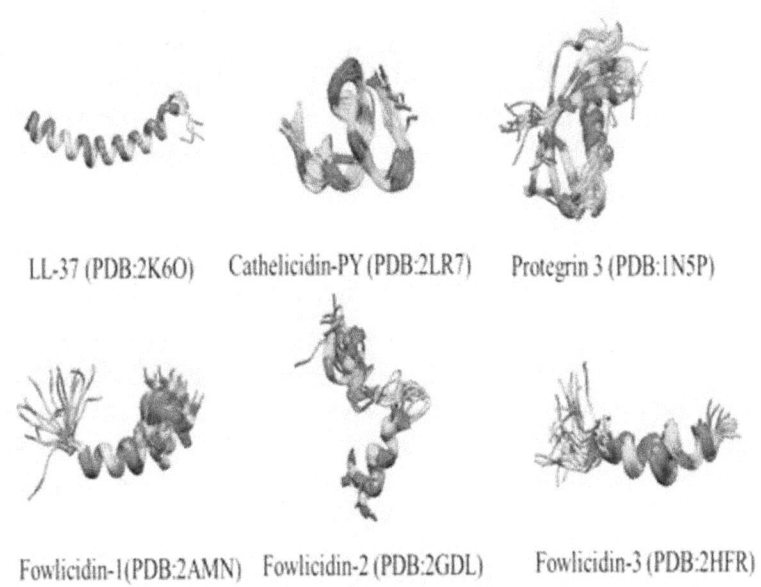

Fig. 3.1.1: Structures of some cathelicidin-derived antimicrobial peptides. These figures are generated by using UCSF Chimera.

The cathelicidin-derived antimicrobial peptide are characterized by the presence of a cathelin like domain on its N terminal and a variable domain on its C terminal. The term cathelicidin was given in 1995 because of presence of cathelin like domain (Lehrer and Ganz 2002). The precursor of cathelicidin-derived preproregion contains 123-144 amino acid residues, which includes a signal peptide of 29-30 amino acid residues (also known as Preregion), followed by a proregion, also called

CHAPTER-3: OVERVIEW ON CATHELICIDIN-DERIVED, PROLINE- RICH AND TRYPTOPHAN-RICH ANTIMICROBIAL PEPTIDES

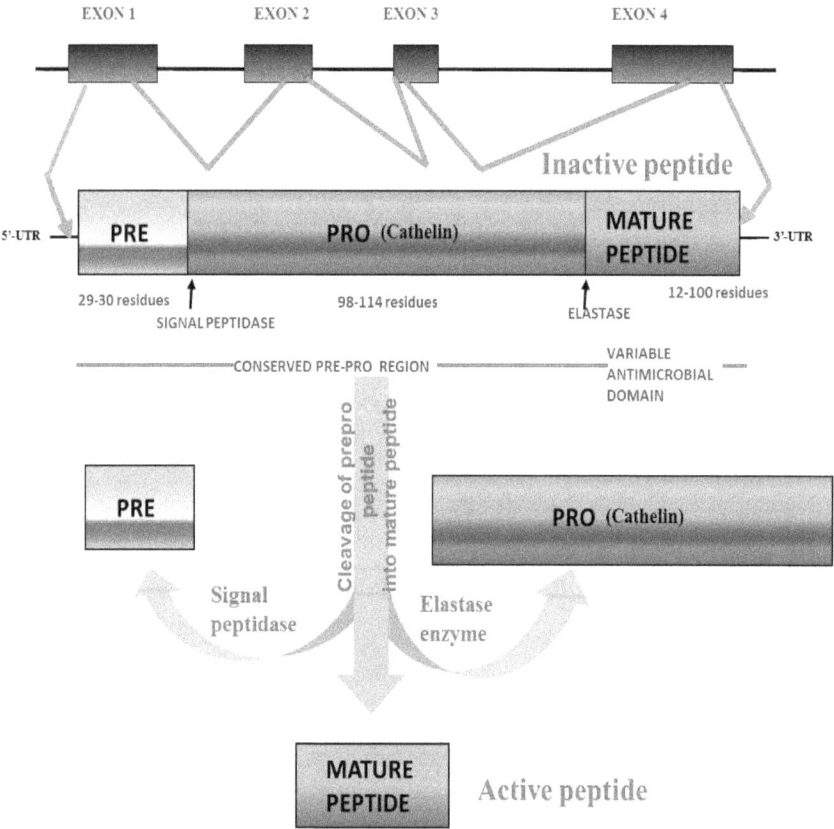

Fig. 3.1.2: A cartoon showing the cleavage process involved in the generation of a mature active antimicrobial peptides from the prepropeptide of a cathelicidin-derived antimicrobial peptide and also the structural organization of their genes. Adopted from (Ahmad et al. 2012).

CHAPTER-3: OVERVIEW ON CATHELICIDIN-DERIVED, PROLINE- RICH AND TRYPTOPHAN-RICH ANTIMICROBIAL PEPTIDES

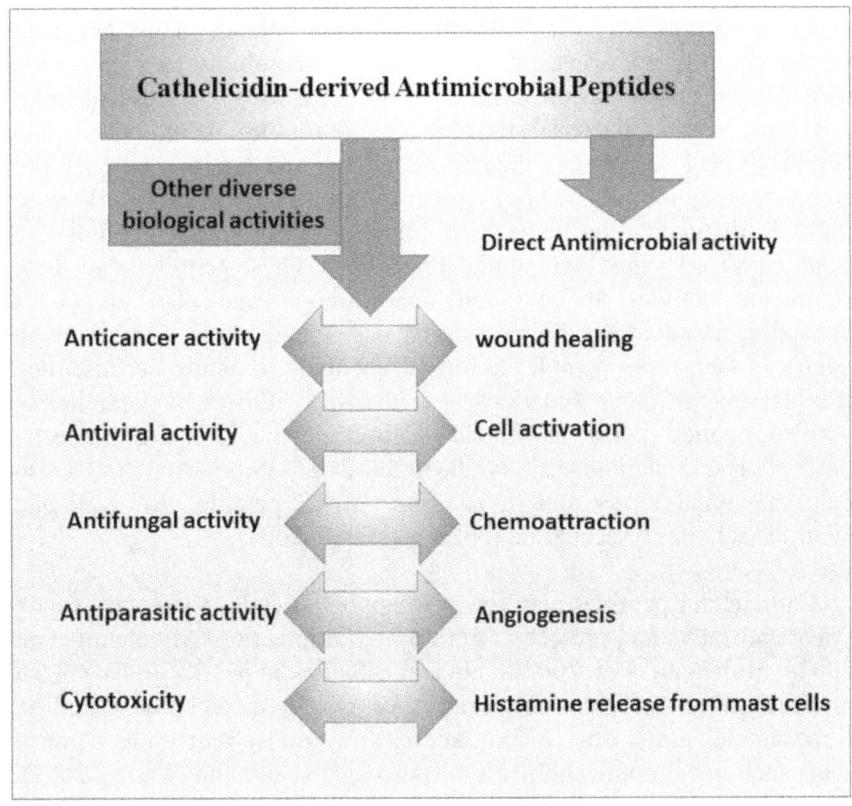

Fig. 3.1.3: Multifunctional role of cathelicidin-derived antimicrobial peptides.

cathelin domain of 94-114 residues (Fig. 3.1.2) and a variable domain of 12-97 residues, which possess antimicrobial property (Zanetti et al. 1995; Gennaro and Zanetti 2000). We know that cathelicidin gene contains four exons and three introns. The fourth exon encode the cleavage site and the varied antimicrobial domain and the preproregion (signal peptide and cathelin-like domain) is encoded by the first three exons (Scocchi et al. 1997; Ramanathan et al. 2002).

CHAPTER-3: OVERVIEW ON CATHELICIDIN-DERIVED, PROLINE- RICH AND TRYPTOPHAN-RICH ANTIMICROBIAL PEPTIDES

Now it is well understood that cathelicidin-derived antimicrobial peptides have a wide range of antimicrobial activity against Gram-negative and Gram-positive bacteria, fungi, parasite and enveloped viruses (Fig. 3.1.3). Some peptides also have the ability to kill bacteria which are resistant to conventional antibiotics. Most of these peptides self-assembled in bacterial membrane and rupture them, but certain peptides translocate across the cell membrane and interact with cellular components to inhibit bacterial growth (Nizet and Gallo 2003) . Because of their distinctive properties and high activity, they are admirable candidates for various therapeutic applications. These peptides also show significant activity at high salt concentration and also absent of disulphide bond make their synthesis easier and cheap (Travis et al. 2000). In fact, these peptides are multifunctional in nature because they exhibit several other functions, for example, PR-39, a cathelicidin derived peptide helps in wound healing in the skin by syndecan production and angiogenesis (Gallo et al. 2002). Similarly, LL-37 also help in wound repair and in addition, LL-37 peptide also bind and neutralizes lipopolysaccharide (Nizet and Gallo 2003).

Cathelicidin peptides also found in epithelial cells, alimentary canal, sweat and saliva and protect us from bacterial infection (Murakami et al. 2002a; Murakami et al. 2002b). Several cathelicidin peptides have shown promising results for possible drug development in preclinical studies, for example, protegrin-1 analog successfully underwent under clinical trials for topical application in oral mucositis (Cole and Waring 2002). Also MBI 594AN is a cathelicidin-based peptide underwent for the treatment of acne and Omiganan (MBI-226), an analog of indolicidin have activity against bacterial and fungal species also underwent clinical trial (Gordon et al. 2005; Hancock and Sahl 2006). Despite these successes, there are many limitations which we have to overcome. Many cathelicidin-derived peptides show good activity against several pathogens, but at same time they are high toxicity to mammalian cells. BMAP-27 and 28 are two well-studied cathelicidin derived peptide which has good antibacterial activities, but simultaneously they are also highly toxic (Ahmad et al. 2009a; Ahmad et al. 2009b). Therefore, we need to work more in this field and find out the factors which are responsible for their high toxicity. Recently, several analogs of these antimicrobial peptides have been designed and characterized which are non-toxic with same antibacterial activity (Ahmad et al. 2009a; Ahmad

CHAPTER-3: OVERVIEW ON CATHELICIDIN-DERIVED, PROLINE- RICH AND TRYPTOPHAN-RICH ANTIMICROBIAL PEPTIDES

et al. 2009b).

3.2 Proline-rich Antimicrobial Peptides

Proline-rich antimicrobial peptides (PR-AMPs) are one of the well-known group of AMPs, characterized by a high content of proline residues and have been isolated from several organisms and have a unique mode of action. It is to be noted that the majority of PR-AMPs are primarily isolated from insects, but it has been also discovered in other organism such as cattle, amphibians, crustaceans, molluscs and mammalian species (Otvos Jr 2002; Scocchi et al. 2011). Drosocin is one of the well-studied shorter size proline rich antimicrobial peptides, isolated from Hymenopetra, Lepidoptera, Hemiptera and Diptera insects (Imler and Bulet 2005). Another peptide known as pyrrhocoricin, which is also shorter in size was discovered in the European sap-sucking bug *Pyrrhocoris apterus* and formaecin was isolated from the bulldog ant *Myrmecia gulosa* (Bulet and Stocklin 2005; Mackintosh et al. 1998). Some small peptides have been also discovered in bees, known as apidaecins (Li et al. 2006). Along with this, some peptides with larger in size was also discovered in insects. For example, metchnikowin from fruit fly (Levashina et al. 1995), heliocins and lebocins from moths, and abaecins from honeybees, were isolated (Bulet and Stocklin 2005). In mammals, PR-AMPs were isolated from pig (PR-39, PRP-SP-8), goat (Bac3.4, Bac 5 and Bac 7), cow (Bac 5, Bac 4 and Bac 7), sheep (Bac 5, Bac 63, Bac7 and Bac 11) and goat (Bac 3.4, Bac 5 and Bac 7) (Otvos Jr 2002; Scocchi et al. 2011). PR-bombesin, a PR-AMPs, discovered in the skin of the toad *Bombina maxima* (Lai et al. 2002). Some PR-AMPs were also isolated from invertebrate, for example, penaeidins from shrimp, arasins and callinectin from crabs, astacidins from crayfish and Cg-PRP from oyster, were isolated (Jiravanichpaisal et al. 2007; Gueguen et al. 2009; Otero-González et al. 2010). Some PR-AMPs are summarized in table-3.2.1.

PR-AMPs usually do not form a helical or β–sheet structure. The presence of proline in the peptide sequence decreases the tendency of a helix formation. These peptides generally form poly-L proline type II helical conformation (PP-II helix), which belongs to extended class of antimicrobial peptides. For instance, apidaecins and drosocin

CHAPTER-3: OVERVIEW ON CATHELICIDIN-DERIVED, PROLINE- RICH AND TRYPTOPHAN-RICH ANTIMICROBIAL PEPTIDES

Table-3.2.1: Proline-rich antimicrobial peptides

Antimicrobial Peptide	Sequence	Source	Reference
Drosocin	GKPRPYSPRPTSHPRPIRV	Fruit fly, *Drosophila melanogaster*	(Bulet et al. 1996)
Pyrrhocoricin	VDKGSYLPRPTPPRPIYNRN	European fire bug *Pyrrhocoris apterus*	(Cociancich et al. 1994a; Cociancich et al. 1994b)
Formaecin 1	GRPNPVNNKPTPHPRL	Red bulldog ant *Myrmecia gulosa*	(Mackintosh et al. 1998)
Apidaecin	GNRPVYIPPPRPPHPRL	*Bombus pascuorum*	(Rees et al. 1997)
Metchnikowin	HRHQGPIFDTRPSPFNPNQPRPGPIY	Common fruit fly, *Drosophila melanogaster*	(Levashina et al. 1995)
Heliocin	QRFIHPTYRPPPQPRRPVIMRA	Owlet moth, *Heliothis virescens*	(Bulet and Stocklin 2005)
Penaeidin-1	YRGGYTGPIPRPPPIGRPPLRLVVCACYRLSVSDARNCCIKFGSCCHLVK	Penoeid shrimp, *Penaeus*	(Destoumieux et al. 2000)

CHAPTER-3: OVERVIEW ON CATHELICIDIN-DERIVED, PROLINE- RICH AND TRYPTOPHAN-RICH ANTIMICROBIAL PEPTIDES

		vannamei	
PR-bombesin	EKKPPRPPQWAVGHFM	Bombina maxima	(Lai et al. 2002)

form the PP-II helix structure (Scocchi et al. 2011). Proline rich bac5 also form PP-II helix structure (Raj and Edgerton 1995) and PR-39 and dipericin also form similar spectra (Otvos Jr 2002).

The majority of AMPs kill the bacteria by interacting with bacterial membrane through electrostatic interaction and kill them through cell lysis. PR-AMPs usually do not interact with bacterial membrane like other antimicrobial peptides. These peptides can cross the bacterial membrane without damaging it and interact with cellular components and thus, stop the bacterial growth. For example, PR-39, penetrates the outer membrane of *E.coli* and stops the protein and DNA synthesis and inhibits the bacterial growth (Boman et al. 1993). These peptides show high specificity toward Gram-negative bacteria and usually do not kill Gram-positive bacteria. It is very much clear that high amounts of negatively charged lipids is present on surfaces of both Gram-negative and Gram-positive bacterial cell walls. Also it is well know that outer layer of Gram-negative bacteria are composed of mainly lipopolysaccharide, a polyanionic molecule whileas outer layer of Gram-positive bacteria are composed of acidic polysaccharides (Teichoic acids) and phosphatidylglycerol (Gutsmann et al. 2001). The charge on both types of bacteria is negative, but their lipid composition is comparatively different. Probably the difference in the lipid composition in the outer leaflet of these bacteria is important for their antibacterial activity. The accumulating data available in the literature suggested that probably several PR-AMPs such as pyrrhocoricin, drosocin and apidaecin bind with the Gram-negative bacterial lipopolysaccharide and enter into bacterial cells and bind with bacterial DnaK (70-kDa heat shock protein) and/or with a GroEL protein (60-kDa bacterial chaperonin) (Otvos et al. 2000; Kragol et al. 2001; Kragol et al. 2002) and in this way, they utilized this type of mechanism to kill specifically Gram-negative bacteria.

CHAPTER-3: OVERVIEW ON CATHELICIDIN-DERIVED, PROLINE- RICH AND TRYPTOPHAN-RICH ANTIMICROBIAL PEPTIDES

3.3 Tryptophan-rich Antimicrobial Peptides

Some antimicrobial peptide contains a high proportion of tryptophan (Trp) residue in their sequence. These antimicrobial peptides are cationic in nature and show wide range of activity towards Gram-positive and Gram-negative bacteria and also against fungi. It is well known that Trp residues have a high preference for the interfacial region of the lipid bilayer (Yau et al. 1998). It has also been observed that Trp residues form hydrogen bonds with water and with components of the lipid bilayer (Lee et al. 2004). The study suggested that the cationic nature of the peptide because of the presence of arginine amino acids, promotes the binding of peptide with membrane and hydrogen binding facilitates its interaction with a negatively charged surface such as lipopolysaccharide, Teichoic acids or phosphatidyl glycerol (Chan et al. 2006). Thus, these peptides are responsible for cationic- π interaction and facilitate peptide membrane interaction. For example, the antimicrobial peptides derived from Lactoferrin, show that the presence of Trp residue in the sequence make the peptide more active (Haug and Svendsen 2001; Haug et al. 2001). Moreover, several Trp rich short size AMPs were designed and characterized, which show that the presence of Trp residue in the sequence, make the peptide more active against bacteria (Strøm et al. 2002). Thus overall, the data available in the literature concluded that the distinct properties of these peptides make them highly active even at very low concentration.

Several peptides such as Indolicidin, tritpticin and puroindoline are Trp-rich antimicrobial peptides (TR-AMPs). Indolicidin is a 13 residue cathelicidin-derived AMPs which show a broad range of activity against bacteria, fungi, protozoa and the enveloped virus human Immunodeficiency virus type 1 (Selsted et al. 1992). This peptide translocates across the outer membrane of bacteria and cause bacterial cell lysis by channel formation (Falla et al. 1996). This peptide does not form regular α-helical or β-sheet structure, but form an extended boat-shaped structure (Rozek et al. 2000a). Tritrpticin is another class of cathelicidin-derived AMPs which are cationic and contain a high proportion of Trp residues. In the presence of Sodium dodecyl sulfate (SDS micelles), this peptide, Tritrpticin, forms an amphipathic conformation.

CHAPTER-3: OVERVIEW ON CATHELICIDIN-DERIVED, PROLINE- RICH AND TRYPTOPHAN-RICH ANTIMICROBIAL PEPTIDES

HIV gp41 (PDB:1JAV)

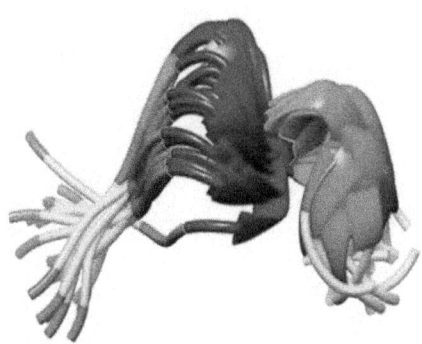

Tritrpticin (PDB:1D6X)

Fig. 3.3.1: Structure of Trp-rich antimicrobial peptides, HIV gp41 and Tritrpticin in the presence of dodecylphosphocholine micelles and sodium dodecyl sulfate micelles respectively. These figures are visualized by using UCSF chimera. The blue color indicates Trp-rich regions.

(Fig. 3.3.1) and forms a wedge shape structure (Schibli et al. 1999). It has also been reported that tritrpticin form ion channel. Puroindoline is

CHAPTER-3: OVERVIEW ON CATHELICIDIN-DERIVED, PROLINE- RICH AND TRYPTOPHAN-RICH ANTIMICROBIAL PEPTIDES

also a Trp rich peptide isolated from wheat seeds. They are found in two isoforms puroindoline-a and puroindoline-b, in which puroindoline-a contain a unique Trp-rich domain (WRWWKWWK). They have the capacity to fight against a pathogen, therefore they are very valuable particularly for the food industry (Capparelli et al. 2005).

Table-3.3.1: Selected examples of Trp-rich antimicrobial peptides.

Antimicrobial Peptide	Sequence	Source	Reference
Indolicidin	ILPWKWPWWPWRR	Bovine neutrophils, *Bos taurus*	(Selsted et al. 1992)
Tritrpticin	VRRFPWWWPFLRR	Porcine neutrophils	(Schibli et al. 1999)
Combi-1	RRWWRF	Synthetic	(Blondelle et al. 1995)
Combi-2	FRWWHR	Synthetic	(Rezansoff et al. 2005)
PAF26	RKKWFW	Synthetic	(López-García et al. 2002)
Pac-525	KWRRWVRWI	Synthetic	(Wei et al. 2006)

These peptide forms well define structure in SDS-micelles which are stabilized by the presence of five disulfide bridges (Le Bihan et al. 1996). It is reported that this peptide also forms an ion channel in biological membrane (Charnet et al. 2003). Several other peptides such as combi-1, combi-2, CP-11, and CP-10A also belong to this class of peptide (Chan et al. 2006). Some AMPs derived from food proteins such as lactoferrin

CHAPTER-3: OVERVIEW ON CATHELICIDIN-DERIVED, PROLINE- RICH AND TRYPTOPHAN-RICH ANTIMICROBIAL PEPTIDES

and lysozyme also belong to Trp-rich peptides, which possess antimicrobial activity (Chan et al. 2006). Along with this, several synthetic TR-AMPs were synthesized and their antibacterial and toxic activities were tested (Strøm et al. 2002; Saravanan et al. 2014).

CHAPTER-4
BIOLOGICAL ACTIVITIES OF ANTIMICROBIAL PEPTIDES

4. BIOLOGICAL ACTIVITIES OF ANTIMICROBIAL PEPTIDES

Antimicrobial peptides are major effector molecules of innate immunity, which are produced by several plants and animals (Boman 2003). They are also known as ancient weapons of plant and animal to fight against bacterial infections (Zasloff 2002). The most interesting thing is that they do not only have antimicrobial activity, but also have several other biological activities (Fig. 4.1) such as anticancer, antiviral and antiparasitic activities (Hancock and Scott 2000). Generally, these peptides interact with the bacterial membranes and formed pores in them by several mechanisms. However, some antimicrobial also killed bacteria by a mechanism similar to conventionally available antibiotics. In this section, we have tried to explain broad spectrum activities of these peptides.

4.1 Antibacterial Activity

It is well documented that antimicrobial peptides have a broad spectrum of activity against several Gram-negative and positive bacteria (Boman 2003). In the lab, several methods are routinely used to check the ability of peptides to inhibit bacterial growth. The antibacterial activity is tested by broth dilution methods or with disc diffusion assay (Andrews 2001). These methods are important to screen the peptides which possess antibacterial activity. In agar or broth dilution methods, we tried to find out the minimum inhibitory concentration (MIC) of peptides (Wiegand et al. 2008). There are also commercial methods for determining MIC, for example, Etest strips and Oxoid MICEvaluator method (Mushtaq et al. 2010). MIC is defined as a concentration of peptides that resulted in

CHAPTER-4: BIOLOGICAL ACTIVITIES OF ANTIMICROBIAL PEPTIDES

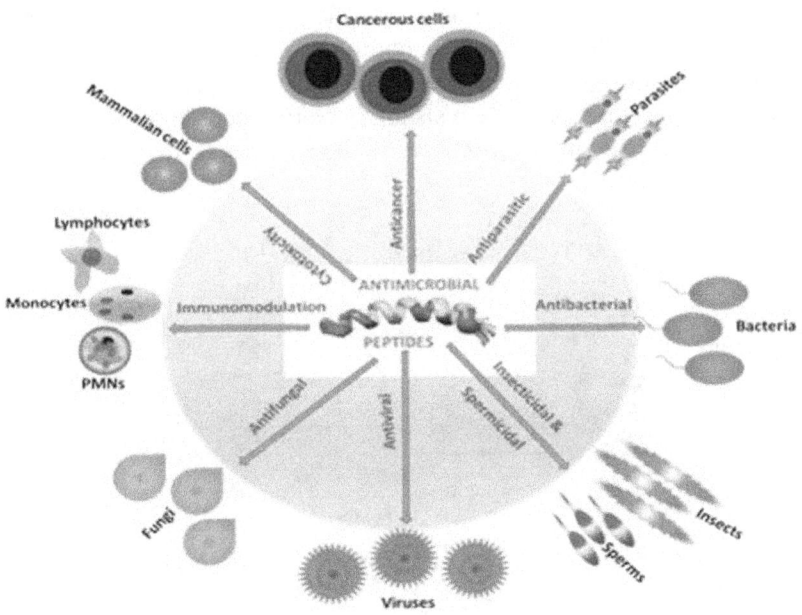

Fig.-4.1: Diverse functions of antimicrobial peptides.

100% inhibition of microbial growth or somewhere it is also defined as the lowest concentration of antimicrobial peptides which inhibit the visible growth of bacterial cells after 16-18 hrs. of incubation at 37°C (Ahmad et al. 2009b; Pandey et al. 2010). MIC is also one of the most basic laboratory methods for determining the activity of an antimicrobial peptides against a bacteria (Ahmad et al. 2009b; Ahmad et al. 2012). Those microorganisms who are able to survive after one or two antibiotic exposures are known as resistant bacteria and those microorganisms who become resistant against multiple antibiotics are called multiple resistant or superbugs. Some antimicrobial peptides are excellent in terms of activity because they are having very good antibacterial activities and show appreciable activity against highly resistant bacteria such as multidrug-resistant *Pseudomonas aeruginosa* (Gram negative bacterium), meticillin-resistant *S. aureus*, (Gram-positive bacterium) and *S. maltophilia.* (Gram-negative bacterium) (Steinberg et al. 1997; Wu and Hancock 1999; Zhang et al. 2000).

CHAPTER-4: BIOLOGICAL ACTIVITIES OF ANTIMICROBIAL PEPTIDES

We know that development of resistant against conventional antibiotic is one of the major problems. Therefore, scientist and researcher all over the world are searching novel antimicrobial agents to overcome this problem. Peptide antibiotic is one option which can be used as a drug against multiple resistant bacteria because of their unique properties which are already discussed in the above text. In our body

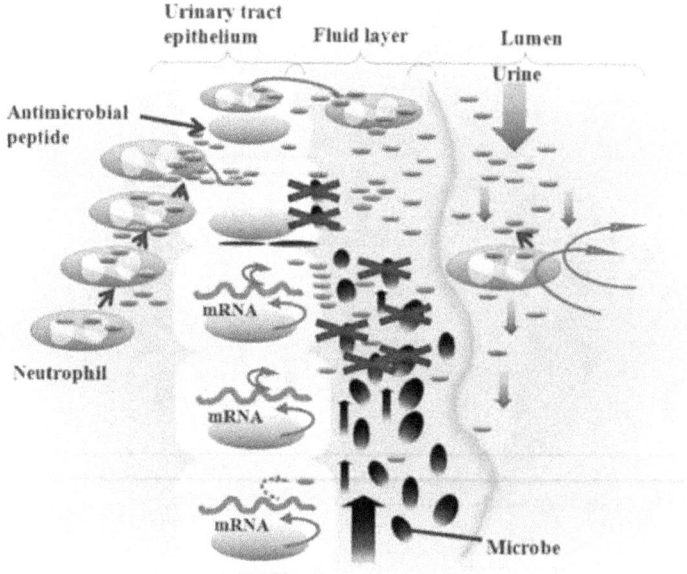

Fig.4.1.1: Representation of biological activity of cathelicidin-derived antimicrobial peptides which protect our urinary tract from bacterial infection.

several antimicrobial peptides are present, which helps in fighting against bacterial infections. The people who lack α-defensins, LL-37 and human beta defensin in their body are more prone to bacterial infections. In psoriasis (Harder and Schröder 2005) and bronchoalveolar inflammation (Hiratsuka et al. 2003; Ross et al. 2004) expression of hBD-2 and hBD-3 are enhanced. It is also well documented that antimicrobial peptides protect our kidney from bacterial infections. Two well-known families of peptides, defensin and cathelicidin are produced in the kidney. These peptides are secreted onto the urinary surface of epithelial cells and show

CHAPTER-4: BIOLOGICAL ACTIVITIES OF ANTIMICROBIAL PEPTIDES

a broad-spectrum of antimicrobial activity and protect our urinary tract from bacterial infection (Fig. 4.1.1) (Zasloff 2006). Shigellosis, (also recognized as bacillary dysentery or Marlow Syndrome), is a type of foodborne illness caused by a group of bacteria called *Shigella*. The sodium butyrate, the salt of a short-chain fatty acid, which is administered by mouth, generally used for stimulating the epithelial lining of the colon and rectum to produce antimicrobial peptides, which kills *Shigella* (Zasloff 2006), and thus, help in treating this disease.

4.2 Anticancer Activity

Cancer is one of the dreadful disease characterized by uncontrolled growth of cells. Cancerous cells do not die and they rapidly continue to grow and divide. Conversely, normal cells undergo programmed cell death, also called apoptosis, and when this process discontinued, the cancer begins to form. The accumulating data suggest that there are more than 200 types of cancer. Furthermore, cancer is one of the major public health problems in the whole world and the possibility of reoccurrence of this disease has made it more dangerous. Cancer can be benign or malignant in nature. Benign tumors are non-cancerous, localized only in a certain region and do not spread to other parts of the body whilst malignant tumors are cancerous, rapidly spread to other parts of body. Surgery is one of the primary

Table-4.2.1: Selected examples of anticancer peptides.

Antimicrobial Peptide	Sequence	Source	Reference
Aurein 1.1	GLFDIIKKIAESI	Frog	(Rozek et al. 2000a)
Melittin	GIGAVLKVLTTGLPALISWI KRKRQQ	Honeybee	(Fennell et al. 1967)
Mastoparan	INLKALAALAKKIL	Insect	(Hirai et al. 1979)
Gomesin	QCRRLCYKQRCVTYCRGR	Spider	(Silva et al. 2000)

CHAPTER-4: BIOLOGICAL ACTIVITIES OF ANTIMICROBIAL PEPTIDES

Maculatin 1.1	GLFVGVLAKVAAHVVPAIAEHF	Frog	(Rozek et al. 1998)
BMP-28	GGLRSLGRKILRAWKKYGPIIVPIIRI	Cattle	(Risso et al. 2002)
BMP-27	GRFKRFRKKFKKLFKKLSPVIPLLH	Cattle	(Risso et al. 1998)
Magainin 2	GIGKFLHSAKKFGKAFVGEIMNS	Frog	(Cruciani et al. 1991; Lehmann et al. 2006)
Brevinin-1EMb	FLPLLAGLAANFLPTIICKISYKC	Frog	(Park et al. 1994)
Aurein 1.2	GLFDIIKKIAESF	Frog	(Rozek et al. 2000b)
Epinicidin-1	GFIFHIIKGLFHAGKMIHGLV	Fish	(Pan et al. 2007)
Tachyplesin I	KWCFRVCYRGICYRRCR	Crustacean hemocytes	(Li et al. 2000)
LL-37/hCAP-18	LLGDFFRKSKEKIGKEFKRIVQRIKDFLRNLVPRTES	Human	(Okumura et al. 2004)

treatment options to treat cancer. Chemotherapy and radiation therapy are other options which are widely used for treating this disease. Most of anticancer drugs are non-selective and show hazards effect on health due to deleterious side effects. Moreover, development of resistant against anticancer drugs is also one of major obstacles to treat this disease. Despite continuous effort in the development of cancer therapy, till date not a single very effective anticancer drug is present in the market.

CHAPTER-4: BIOLOGICAL ACTIVITIES OF ANTIMICROBIAL PEPTIDES

Therefore, new anticancer drug with high selectivity and lower mammalian toxicity is urgently required. The peptide based anticancer drug is one of the hopeful approach for treating this malady. In this section, we will discussed about the naturally occurring anticancer peptides and also several approaches, which is usually used for designing anticancer peptides.

Membranes of mammalian cells are neutral in nature, because they are composed of zwitterionic lipids, whereas bacterial membranes are negatively charged because of the presence of anionic lipid. Similarly, cancer cells are different from normal cells in terms of lipid composition. Membranes of cancer cells are negatively charged because of the presence of anionic lipid. The cancerous cell contains Phosphatidylserine in outer membrane (Zwaal et al. 2005; Zwaal and Schroit 1997) and also concentration of sialic acid increases in these cells, which provide overall negative charge, makes these cells different from normal cells. Antimicrobial peptides are usually cationic and therefore due to difference in charge between normal and the cancerous cell line, they specifically target the cancer cell line through electrostatic interaction. They kill the cancerous cell within minutes and therefore does not leave any possibility of developing resistance. Numerous types of anticancer cationic peptides have been reported in the literature (Hoskin and Ramamoorthy 2008). These peptides also translocate into cytsol and interact preferentially with anionic lipid of mitochondrial membrane and may stimulate apoptosis, and cell death (Ellerby et al. 1999). It is also reported that several peptides caused cancer cell death through necrosis. Several naturally occurring antimicrobial peptides were isolated from different organism which exhibits strong antitumor activity, for example, BMAP-27 and BMAP-28 which were isolated from cattle, depicted appreciable anticancer activity (Risso et al. 1998; Risso et al. 2002). A cathelicidin-derived antimicrobial peptide, isolated from human body also show remarkable anticancer activities (Okumura et al. 2004). Cecropins, a group of antimicrobial peptides found in insects, silk moth Hyalophora cecropia and mammals, also exhibit strong antitumor activity with low mammalian cell toxicity (Hoskin and Ramamoorthy 2008). Many antimicrobial peptides discovered in amphibians such as aurein 1.2, citropin 1.1, gaegurins, and magainins are toxic to cancer cells (Hoskin and Ramamoorthy 2008). A well-known naturally occurring antimicrobial peptide, melittin isolated from honey bees, also reported to be have strong cytotoxic activity

CHAPTER-4: BIOLOGICAL ACTIVITIES OF ANTIMICROBIAL PEPTIDES

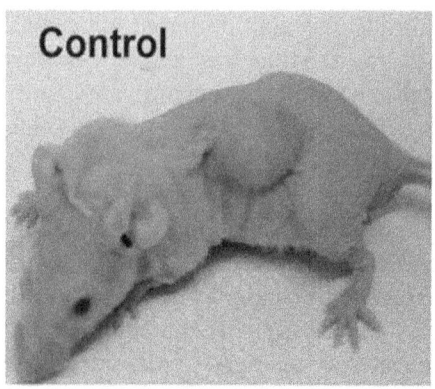
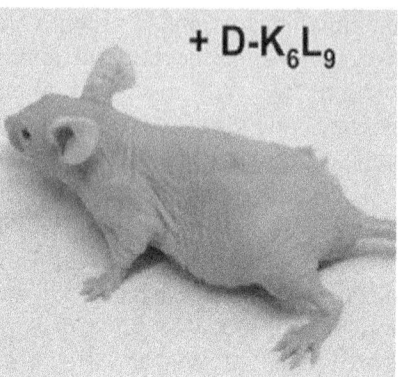

Figure-4.2.1: Effect of anticancer peptide on tumor. The above figure shows reduction in tumor size when mice was administrated with anticancer peptide (D-K 6 L 9). Left side of the figure show male mice treatment without peptide and right side of the figure shows treatment with peptide. This figure is adopted from (Papo et al. 2006).

against cancerous cells (Fennell et al. 1967). Other antimicrobial peptides like defensins, lactoferricin, Tachyplesin I and linear peptide PR-39 which form significant beta sheet structure, show good anticancer activities (Hoskin and Ramamoorthy 2008).

In recent years, several antimicrobial peptides were designed and characterized which show appreciable anticancer activities. Among these, diastereomeric amino acids containing peptides are of great interest, for example, D- K4R2L9 is a diastereomeric 15 residue amphipathic peptide composed of leucines, arginines, and lysines, residues selectively kill cancerous cells without showing any cellular toxicity to normal cells (Papo et al. 2003; Papo and Shai 2003b). Another anticancer peptide, D-K6L9, a 15-amino acid diastereomeric amphipathic peptide (Fig.-4.2.1), show selective in vitro cytotoxic activity against human prostate cancer cell lines (Papo et al. 2006; Gaspar et al. 2014). Moreover, another peptide containing only L amino acids, L-K6L9 were designed which show similar activity, but highly toxic to human red blood cells (Papo et al. 2004).

Several peptides were designed either by substitution, or deletion or

CHAPTER-4: BIOLOGICAL ACTIVITIES OF ANTIMICROBIAL PEPTIDES

by changing key amino acids in naturally occurring antimicrobial and some anticancer peptides were rationally designed based on the structure function relationship of antimicrobial peptides. A peptide called KLAK, an antimicrobial proapoptotic peptide, were designed and characterized which show good anticancer activities both in vitro and in vivo (Ellerby et al. 1999; Mai et al. 2001). This peptide is coupled with BiP peptide separated by diglyicine spacer, to produce the BiP/KLAK peptide,which has the ability to kill cancer cells (Obeid 2009). Moreover, this peptide is also coupled with PTD peptide, separated by diglyicinespacer, to generate the proapoptotic peptide DP1 could be functional in vitro use and for intratumor injection (Mai et al. 2001). Some analogs of temporin-1CEa were rationally designed and synthesized which show improved anticancer activity (Yang et al. 2013). Some antimicrobial peptides which act as an anticancer agent are summarized in table-4.2.1.

4.3 Antifungal Activity

A fungus is a group of eukaryotic organisms, including yeasts, molds and mushroom, which are classified in a group called kingdom Fungi. The science which deals with the study of fungi is called mycology. Several drugs are produced from fungi which have various medicinal uses. Moreover, many fungi are edible and some are highly poisonous. Some fungus produces a biologically active compound, which are toxic to both animals

Table-4.3.1: Selected examples of antifungal peptides.

Antimicrobial Peptide	Sequence	Source	Reference
Drosomycin	DCLSGRYKGPCAVWDNETCRRVCKEEGRSSGHCSPSLKCWCEGC	Fruitfly	(Fehlbaum et al. 1994)
Cecropin P2	SWLSKTYKKLENSAKKRISEGIAIAIQGGPR	Nematode	(Pillai et al. 2005)
Heliomicin	DKLIGSCVWGAVNYTSDCNGECKRRGYKGGHCGSFANVNCWCET	Tobacco budworm	(Lamberty et al. 2001)

CHAPTER-4: BIOLOGICAL ACTIVITIES OF ANTIMICROBIAL PEPTIDES

Maximin 1	GIGTKILGGVKTALKGALKELASTYAN	Toad	(Lai et al. 2002)
Brevinin-1E	FLPLLAGLAANFLPKIFCKITRKC	Frog	(Simmaco et al. 1994)
Gaegurin-3	GIMSIVKDVAKTAAKEAAKGALSTLSCKLAKTC	Frog	(Park et al. 1994)
Cecropin A	RWKVFKKIEKVGRNIRDGVIKAAPAIEVLGQAKAL	Tobacco budworm moth	(DeLucca et al. 1997)
Temporin L	FVQWFSKFLGRIL	Frog	(Simmaco et al. 1996)
Magainin 2	GIGKFLHSAKKFGKAFVGEIMNS	Frog	(Zasloff 1987)
Hepcidin 20	ICIFCCGCCHRSKCGMCCKT	Human	(Park et al. 2001)
Circulin A	GIPCGESCVWIPCISAALGCSCKNKVCYRN	*Chassalia parviflora*	(Fujikawa et al. 1965)
LL-37	LLGDFFRKSKEKIGKEFKRIVQRIKDFLRNLVPRTES	Human	(Agerberth et al. 1995)
BMAP-27	GRFKRFRKKFKKLFKKLSPVIPLLHLG	Cattle	(Skerlavaj et al. 1996)
BMAP-28	GGLRSLGRKILRAWKKYGPIIVPIIRIG	Cattle	(Skerlavaj et al. 1996)

CHAPTER-4: BIOLOGICAL ACTIVITIES OF ANTIMICROBIAL PEPTIDES

Piscidin 1	FFHHIFRGIVHVGKTIHRLVTG	Fish	(Silphadung and Noga 2002)
Lycotoxin I	IWLTALKFLGKHAAKHLAKQQLSKL	Spider	(Yan and Adams 1998)
Sesquin	KTCENLADTY	Plants	(Wong and Ng 2005)
Human Histatin 1	DSHEKRHHGYRRKFHEKHHSHREFPFYGDYGSNYLYDN	Human	(Oppenheim et al. 1988)

and plants and are therefore known as mycotoxins. Fungal disease is always very dangerous and often caused by fungi which are found in the environment. They are generally found in soil, plants, trees, and animals including human beings. In recent years, there is a remarkable rise in the incidence and diversity of fungal infections in human beings. The demand of antimicrobial peptides has increased rapidly because of toxicity and resistance against currently employed antifungal agents. Some antimicrobial peptides represent potential candidates to kill the fungal pathogen that is a serious menace to human health (van der Weerden et al. 2013). To design novel antifungal peptides, understanding of the mechanism of action of antifungal peptides against fungus is urgently required.

A plenty of antifungal peptides have been discovered in several microorganisms and also in bacteria, plants and animals. The first antifungal peptides, iturin and bacillomycin families, have been isolated from *Bacillus subtilis*. Later on, syringomycins, syringostatins, and the syringotoxins have been isolated from *Pseudomonas syringae* and their antifungal properties were checked and it has been found that they are lethal to *Aspergillus flavus, Aspergillus fumigatus, Aspergillus niger, Fusarium moniliforme, and Fusarium oxysporum and Candida albicans*. Cepacidines and nikkomycins antifungal agents are found in

CHAPTER-4: BIOLOGICAL ACTIVITIES OF ANTIMICROBIAL PEPTIDES

Burkholderia cepacia and *Streptomyces tendae* respectively. A large number of antifungal peptides have been also isolated from plants such as zeamatin, amphibine H, frangufoline, nummularine, and rugosanine A. Several antifungal peptide have been isolated from various fungal species which are toxic to many pathogenic fungal species. Echinocandins, isolated from *Anidulans and Aspergillus rugulosus* and pneumocandins, isolated from *Zalerion arboicola*, are effective antifungal peptides but also highly toxic to mammalian cells. The aureobasidins, an antifungal peptide, isolated from *Aureobasidium pullulans,* show potent antifungal activity and along with this leucinostatins and helioferins is a family of potent antifungal peptides but unfortunately they are also highly toxic to mammalian cells. Many antifungal peptides have been also discovered in insects Cecropins are a group of lytic peptides produced by the giant silk moth, Hyalopora cecropia which are lethal against fungal species. Moreover, drosomycin and thanatin are cysteine-rich antifungal peptides, produced by *Drosophila melanogaster* and *Podisus maculiveris*, respectively (Fehlbaum et al. 1994; De Lucca and Walsh 1999). Amphibian antifungal peptides have been also isolated such as dermaseptins and magainin, which have potent antifungal activities (De Lucca et al. 1998; Zasloff 1987). Mammalian antimicrobial peptides, α-defensins and β-defensins, mainly produced by human epithelial cells are also lethal against fungal species. Moreover, histatins antimicrobial peptides isolated from human saliva also having an antifungal activity and also analogs have been designed which show improved antifungal activities (Helmerhorst et al. 1997).

As mentioned in the above text that a large number of antimicrobial peptides showed activity against several species of fungus which are harmful for human beings. We have seen that in last a few years the research in antifungal peptides has been increased tremendously. Accumulating data suggested that several designed peptides showed promising antifungal activity. There is not any conserved domain which is responsible for antifungal domain. It is to be noted that modification of naturally occurring antimicrobial peptides is supposed to improve antifungal activity. The data available in the literature showed that conjugation of fatty acid with antifungal peptide has remarkably increased the antifungal activity, for example, conjugation of magainin to undecanoic acid or palmitic, improved their antifungal activity (Avrahami and Shai 2003). In 2006, Yechiel Shai group has synthesized

CHAPTER-4: BIOLOGICAL ACTIVITIES OF ANTIMICROBIAL PEPTIDES

several ultrashort lipopeptides which show efficient antifungal activity (Makovitzki et al. 2006). The antifungal activity could be also improved by conjugating magainin 2 and cecropin A (Resultant peptide is known as P18) (Lee et al. 2004).

How these antimicrobial peptides interact with fungus and caused their lysis is not well understood. However, data available till date suggest that most of the antifungal peptides kill the fungus by forming pores in the membrane or through interaction with cell wall synthesis. Moreover, some peptides interfere with the biosynthesis of essential cellular components such as glucan or chitin to inhibit the growth of fungus (Debono and Gordee 1994). Echinocandins, Pneumocandins, Aculeacins, Mulundocandins, and Aureobasidins are examples of some antifungal peptides which affect glucan synthesis to control fungus growth. Some families of antifungal peptides, for example, Nikkomycins, isolated from *Streptomyces tendae* and polyoxins, isolated from *Streptomyces cacaoi* inhibit chitin synthase and cause fungal cell death (Chapman et al. 1992; Hori et al. 1974). Indolicidin caused fungal cell death by rupturing the membrane organization of fungus (Lee et al. 2003). It has been also reported that some antimicrobial peptides like histatin 5 and lactoferrin-derived peptides exert their fungicidal mechanism through the formation of reactive oxygen species (Jenssen et al. 2006). A few antimicrobial peptides which are used as antifungal agent are summarized in table-4.3.1.

4.4 Antiviral Activity

Viruses are very tiny infectious agent, which can replicate only inside the organism. Viruses can infect several organism, including plants, animals and human beings. These viruses cause several diseases such as the common cold, flu (influenza), smallpox, chickenpox, and warts and also cause many serious diseases, including human immunodeficiency virus infection/acquired immunodeficiency syndrome (HIV/AIDS), avian influenza (avian fluorbird flu), severe acute respiratory syndrome (SARS) and hemorrhagic fevers (Also called Ebola virus disease, EVD) or Ebola hemorrhagic fever, EHF). As we know that viruses multiply inside the host and therefore it is very difficult to treat viral disease. Vaccination is one of the cheapest and effective way to preventing infections caused by viruses. Use of vaccination has dramatically decreased the morbidity and mortality of some well-known viral infected diseases such as polio, measles, rubella, smallpox and mumps (Asaria

CHAPTER-4: BIOLOGICAL ACTIVITIES OF ANTIMICROBIAL PEPTIDES

and MacMahon 2006; Lane 2006). Antiviral drugs inhibit the development of viruses and therefore, used specifically for treating viral infections (Daniels and Nicoll 2011).

Despite extensive studies on the development of antiviral drugs from last several years only a handful of effective antiviral drugs are available in the market. That is why many scientists in the world are working to investigate new candidates which have antiviral activity with distinct modes of action. Some antimicrobial peptides possess good antiviral activity and have the features for the development of peptide based antiviral drugs. It is well established that several antimicrobial peptides have wide and diverse activity against many types of viruses, for instance, defensins, a group of human antimicrobial peptides, inhibits viral growth by interacting with the viral envelope (Klotman and Chang 2006). Another group of antimicrobial peptides called cathelicidin like LL-37, isolated from human beings and indolicidin, isolated from bovine neutrophils, also efficiently inhibits viral replication (Howell et al. 2004; Robinson et al. 1998). Some antimicrobial peptides isolated from amphibians, including dermaseptin S4, caerin 1.1, caerin 1.9, Maximin, Brevinin-1, Temporin A and maculatin 1.1 show appreciable antiviral activities (Lorin et al. 2005; VanCompernolle et al. 2005). Several antimicrobial peptides isolated from insects such as Cecropin A, melittin, ponericin L2, Melectin (Steinberg et al. 1997; Strominger 2009; Wachinger et al. 1998; Orivel et al. 2001; Moreno-Habel et al. 2012; Čeřovský et al. 2008) have good antiviral activities. One of the well-known peptide, plectasin, has been discovered in fungi (Mygind et al. 2005), having a sufficient antifungal activity and show broad spectrum activity against several bacteria which are resistant against conventional

Table-4.4.1: Selected examples of antiviral peptides.

Antimicrobial Peptide	Sequence	Source	Reference
Temporin A	FLPLIGRVLSGIL	Frog	(Simmaco et al. 1996)
Aurein 1.2	GLFDIIKKIAESF	Frog	(Rozek et

CHAPTER-4: BIOLOGICAL ACTIVITIES OF ANTIMICROBIAL PEPTIDES

			al. 2000b)
Melittin	GIGAVLKVLTTGLPALISWIKRKRQQ	Honeybee	(Fennell et al. 1967)
Indolicidin	ILPWKWPWWPWRR	Cattle	(Selsted et al. 1992)
Polyphemusin I	RRWCFRVCYRGFCYRKCR	Arthropod	(Miyata et al. 1989)
Tachyplesin I	KWCFRVCYRGICYRRCR	Arthropod	(Li et al. 2000)
BMP-28	GGLRSLGRKILRAWKKYGPIIVPIIRI	Cattle	(Risso et al. 2002)
BMP-27	GRFKRFRKKFKKLFKKLSPVIPLLH	Cattle	(Risso et al. 1998)
Magainin 2	GIGKFLHSAKKFGKAFVGEIMNS	Frog	(Cruciani et al. 1991)
Plectasin	GFGCNGPWDEDDMQCHNHCKSIKGYKGGYCAKGGFVCKCY	Fungi	(Mygind et al. 2005)
Gramicidin A	VGALAVVVWLWLWLW	Bacteria	(Dubos 1939)
Melectin	GFLSILKKVLPKVMAHMK	Arthropod	(Čeřovský et al. 2008)

antibiotics. The study also revealed that they have the ability to cure peritonitis and pneumonia in mice caused by *S. pneumonia* infection and

CHAPTER-4: BIOLOGICAL ACTIVITIES OF ANTIMICROBIAL PEPTIDES

also shows very low toxicity (Mygind et al. 2005). Piscidin 1, Piscidin 2, Piscidin 3, hepcidin TH1-5, and Epinecidin-1 have been found in fish and reported to have antiviral activities (Wang et al. 2010a; Chinchar et al. 2004; Wang et al. 2010b). Plants are also a rich source of antiviral peptides from which several antiviral peptides have been identified, such as Sesquin, Ginkbilobin, Kalata B8, Cycloviolins A-D, Cycloviolacin Y1, Cycloviolacin Y4, Cycloviolacin Y5 and Circulin C-F (Gustafson et al. 2000; Hallock et al. 2000; Wang et al. 2007; Wang et al. 2008; Daly et al. 2006; Conlan et al. 2011). A few antimicrobial peptides which are used as antiviral agent are summarized in table-4.4.1.

Several attempts have been made to improve the antiviral activity of existing naturally occurring peptides and also some synthetic antiviral peptides have been designed. It is very much clear that antimicrobial peptides with high cationic charge and amphiphilic in nature possess good antiviral activity. In 2004, Gutteberg group has developed several highly cationic peptides which exhibit high antiviral activity (Jenssen et al. 2004). Fragments of antimicrobial peptides have been also created which show improved antimicrobial activity, for example, FK-13 a shorter version of LL-37 cathelicidin derived peptides is active against human immunodeficiency virus (HIV) (Wang et al. 2008) and also several shorter versions of BMAP-27, a cathelicidin derived peptides is active against human immunodeficiency virus (HIV). A shorter version of BMAP-27 known as BMAP-18 has been manufactured in which two phenylalanines were incorporated in place of isoleucine/leucine residues, and the resultant peptide was inactive, indicating the importance of aromatic rings for antiviral activity (Wang et al. 2008). The introduction of proline residue in the sequence of BMAP-18 decreases the antiviral activity because probably distortion of the helical structure of the peptide (Skerlavaj et al. 1996; Wang et al. 2008). Taken together, the data suggested that sequence of peptides, secondary structures, and aromatic residues are important for HIV inhibition. The hydrophobicity of antiviral peptides also play an important role in determining the antiviral activity, for example, in cecropin A (1–8)-magainin 2 (1–12) (termed CA-MA) hybrid peptide, the substitution of serine amino acids with a hydrophobic amino acids significantly increases the antiviral activity (Lee et al. 2004). Lactoferricin and polyphemusin antimicrobial peptides have β structures stabilized by internal disulfide bridges, and it is believed that the presence of these disulfide bridges is accountable for the antiviral activity of these peptides (Jenssen et al. 2004; Jenssen

CHAPTER-4: BIOLOGICAL ACTIVITIES OF ANTIMICROBIAL PEPTIDES

2005). It is widely believed that generally antiviral peptides interact with the envelope of viruses and inhibit their replication.

4.5 Antiparasitic Activity

Parasites are those organisms that live in another organism, known as host and in most of the cases they cannot live without a host. They entirely depend on host for survival and it is believed that they use host to live, grow and multiply. There are numerous types of parasites such as ectoparasite, Endoparasite, Epiparasite and Parasitoid. The ectoparasite lives on surface of host, for example, some mites and lice of hair and body. The Endoparasite lives inside the host such as tapeworm, flukes and roundworms. Ectoparasite are that parasites who live on another parasite for survival. Parasitoids are those organism who lives a significant portion of their life on the host and ultimately kill the host or consume them. Most of these parasites are very harmful to human being and cause several diseases known as parasitic diseases. Many parasites directly affect human beings, but some parasites they release a toxin into the body. Most important parasites causing disease are protozoa (For example, *Entamoeba)*, helminths (For example, tapeworm, flukes and roundworms) and ectoparasites.

Several familiar diseases such as leishmaniasis, trypanosomiasis, and schistosomiasis are caused by parasites. The accumulating data suggest that these diseases are very unsafe for human health. We know

Table-4.5.1: Selected examples of antiparasitic peptides.

Antimicrobial Peptide	Sequence	Source	Reference
Melittin	GIGAVLKVLTTGLPALISWIKRKRQQ	Honeybee	(Fennell et al. 1967)
BMP-28	GGLRSLGRKILRAWKKYGPIIVPIIRI	Cattle	(Risso et al. 2002)
BMP-27	GRFKRFRKKFKKLFKKLSPVIPLLH	Cattle	(Risso et al. 1998)

CHAPTER-4: BIOLOGICAL ACTIVITIES OF ANTIMICROBIAL PEPTIDES

Gomesin	QCRRLCYKQRCVTYCRGR	Spider	(Silva et al. 2000)
Magainin 2	GIGKFLHSAKKFGKAFVGEIMNS	Frog	(Cruciani et al. 1991)
LL-37/hCAP-18	LLGDFFRKSKEKIGKEFKRIVQRIKDFLRNLVPRTES	Human	(Okumura et al. 2004)
Cecropin A	KWKLFKKIEKVGQNIRDGIIKAGPAVAVVGQATQIAK	Insect	(Stciner et al. 1981)
Andropin	VFIDILDKMENAIHKAAQAGIGIAKPIEKMILPK	Fruit fly	(Date-Ito et al. 2002)
Decoralin	SLLSLIRKLIT	Insect	(Konno et al. 2007)
Scorpine	GWINEEKIQKKIDERMGNTVLGGMAKAIVHKMAKNEFQCMANMDMLGNCEKHCQTSGEKGYCHGTKCKCGTPLSY	scorpions	(Conde et al. 2000)
Bicarinalin	KIKIPWGKVKDFLVGGMKAV	Ant	(Rifflet et al. 2012)
Alamethicin	PAAAAQAVAGLAPVAAEQ	Fungi	(Meyer and Reusser 1967)
Gambicin	MVFAYAPTCARCKSIGARYCGYGYLNRKGVSCDGQTTINSCEDCKRKFGRCSDGFITECFL	Insect	(Vizioli et al. 2001)

CHAPTER-4: BIOLOGICAL ACTIVITIES OF ANTIMICROBIAL PEPTIDES

| Micrococcin P1 | SCTTCVCTCSCCTT | Bacteria | (Bennallack et al. 2014) |

that resistant against antibiotic is increasing day by day, therefore we need novel effective therapeutic agents to control these diseases. Some antimicrobial peptides possess antileishmanials activities such as temporin, dermaseptins, gomesin, indolicidin, cecropins, tachyplesin, BMAP-28, Andropin, and also some synthetic peptides, cecropin A-melittin hybrid peptide, BMAP-27 analog, (Akuffo et al. 1998; Feder et al. 2000; Vouldoukis et al. 1996; Bera et al. 2003; Mangoni et al. 2005; Haines et al. 2009; Pérez-Cordero et al. 2011; Lynn et al. 2011). The accumulating data suggested that these peptides primarily act on the membrane and destroy the whole integrity of parasite and leave very little scope for developing resistance. Moreover, peptides such as tachyplesin, magainin analogs, dermaseptins, and phylloseptins show Trypanocidal activities (Huang et al. 1990; Brand et al. 2002; Leite et al. 2005). Many antimicrobial peptides showed remarkable antimalarial activities, for example, cecropin, Meucin-24, magainin, and synthetic hybrids of cecropin and melittin, which are the linear amphipatic peptides are active against Plasmodium species (Gao et al. 2010). Along with this, cecropin-like synthetic peptide, Shiva-3 and dermaseptin analogs show significant activity against Plasmodium species. Several cysteine-rich cationic peptides, defensins, gambicin and scorpine also show inhibitory activity against Plasmodium species (Vizioli and Salzet 2002).

4.6 Insecticidal and Spermicidal Activities

Some insects are very harmful for the agriculture and also cause diseases in human beings. Therefore, several industries have been set to increase the production of pesticide to fulfil the demand. We know that the substance which is used for killing insects are known as insecticide or pesticide, which is a chemical or biological agent (Table-4.6.1). It has also been observed in the last few years that pesticide resistant is increasing day by day, which is an alarming signal for a researcher and scientist to produce distinct insecticides to replace currently available pesticide.

CHAPTER-4: BIOLOGICAL ACTIVITIES OF ANTIMICROBIAL PEPTIDES

Table-4.6.1: Selected examples of insecticidal peptides.

Antimicrobial Peptide	Sequence	Source	Reference
Lycotoxin I	IWLTALKFLGKHAAKHLAKQQLSKL	Spider	(Yan and Adams 1998)
Cupiennin 1a	GFGALFKFLAKKVAKTVAKQAAKQGAKYVVNKQME	Spider	(Kuhn-Nentwig et al. 2002)
Ponericin G1	GWKDWAKKAGGWLKKKGPGMAKAALKAAMQ	Ants	(Orivel et al. 2001)
Esculentin-1	GIFSKLGRKKIKNLLISGLKNVGKEVGMDVVRTGIDIAGCKIKGEC	Frog	(Simmaco et al. 1993)
Cycloviolacin O1	GIPCAESCVYIPCTVTALLGCSCSNRVCYN	Plant	(Craik et al. 1999)

Table-4.6.2: Selected examples of spermicidal peptides.

Antimicrobial Peptide	Sequence	Source	Reference
LL-37	LLGDFFRKSKEKIGKEFKRIVQRIKDFLRNLVPRTES	Human	(Agerberth et al. 1995)
Nisin A	ITSISLCTPGCKTGALMGCNMKTATCHCSIHVSK	Bacteria	(Rogers 1928)
Dermaseptin-	ALWKTMLKKLGTMALHA	Frog	(Mor et al.

CHAPTER-4: BIOLOGICAL ACTIVITIES OF ANTIMICROBIAL PEPTIDES

S1	GKAALGAAADTISQGTQ		1991)
Magainin 2	GIGKFLHSAKKFGKAFVGEIMNS	Frog	(Zasloff 1987)
Maximin 3	GIGGKILSGLKTALKGAAKELASTYLH	Toad	(Lai et al. 2002)
Gramicidin A	VGALAVVVWLWLWLW	Bacteria	(Dubos 1939)
Pediocin PA-1	KYYGNGVTCGKHSCSVDWGKATTCIINNGAMAWATGGHQGNHKC	Bacteria	(Henderson et al. 1992)
Subtilosin A	NKGCATCSIGAACLVDGPIPDFEIAGATGLFGLWG	Bacteria	(BABASAKI et al. 1985)

Some antimicrobial peptides have the ability to control the insects and because of this they are known as insecticidal antimicrobial peptides. VrCRP (V. radiata defensin 1), a plant defensin peptide, isolated from a bruchid-resistant mungbean, exhibits insecticidal activity against *C. chinensisin* (Chen et al. 2004). The cyclotides antimicrobial peptides, kalata B1and B2, Parigidin-br1, and cycloviolacin O1, isolated from plants exhibit insecticidal activity (Plan et al. 2008; Craik 2012; Pinto et al. 2012). A family of antimicrobial peptides named ponericins, showing insecticidal properties, was isolated from the venom of the predatory ant Pachycondyla goeldii (Orivel et al. 2001). Lycotoxins and cupiennin 1 are insecticidal antimicrobial peptides, isolated from spiders ((Yan and Adams 1998; Kuhn-Nentwig et al. 2002). Furthermore, an amphibian antimicrobial peptide, Esculentin-1, also shows insecticidal property ((Ponti et al. 2003). A list of insecticidal peptides has been summarized in table-4.6.1.

Spermicide is a contraceptive substance which has the capacity to abolish sperm. They can be used as alone for contraceptive purpose or they can be used in combination with other methods for example, condoms. Nonoxynol-9, and Menfegol are two examples of spermicide which are available in the market. These available spermicide products have several side effects such as irritation and itching (Xu et al. 2003). Recently it has been discovered that several antimicrobial peptides possess spermicidal properties and they could be considered as a commercial contraceptive agent. Subtilosin A, isolated from *Bacillus subtilis,* have spermicidal property. It is to be noted that nisin, isolated from *Lactococcus lactis,* also has spermicidal property already approved by the Food and Drug Administration (FDA) (Sutyak et al. 2008). Maximin 1 and Maximin 3, isolated from the Chinese red belly toad, also has a spermicidal effect (Lai et al. 2002; Clara et al. 2004). The experiment carried out in male and female bonnet monkeys with magainin-A showed that it can cause complete sperm immobilization and it is an effective and safe intravaginal contraceptive substance (Clara et al. 2004). The study also revealed that dermaseptin (DS1 and DS4) is an effective agent to kill sperm (Zairi et al. 2005). A list of spermicidal peptides has been summarized in table-4.6.2.

4.7 Immunomodulation

Regulation of immune system is called immunomodulation and substance that helps in regulation of the immune system is called Immunomodulators (Dhama et al. 2015). Induction, suppression, or enhancement of the immune response is done by using Immunomodulator for treating disease is known as immunotherapy (Weir et al. 2011). The immunotherapy based on amplification of immune response is called activation immunotherapy whilst immunotherapy based on suppression of immune of response is called suppression immunotherapy. Several synthetic, recombinant or natural molecules are used as immunomodulators (Kayser et al. 2003). Many antimicrobial peptides also have excellent immunomodulatory properties and it has been seen from last a few years interest in this field has tremendously increased because of their remarkable role in both clearance of pathogens and subsequent wound healing healing (Bowdish et al. 2006). Some selected examples of antimicrobial immunomodulatory functions include induction in chemokine production , chemotaxis of lymphocytes, inhibiting cytokine production, promoting wound healing and angiogenesis, modulating the responses of dendritic cells and can also stimulate mast cell degranulation (Bowdish et al. 2005).

It is to be mentioned that several antimicrobial peptides promote or

CHAPTER-4: BIOLOGICAL ACTIVITIES OF ANTIMICROBIAL PEPTIDES

inhibit or complement many cellular functions including chemotaxis, gene transcription, apoptosis, cytokine production and wound healing (Vizioli and Salzet 2002; Salzet 2002; Finlay and Hancock 2004; Mookherjee et al. 2006). For example, LL-37, is a potent immunomodulatory antimicrobial peptide, expressed in high concentration at epithelial surfaces of the human body during bacterial infections. The high concentration of this peptides induce epithelial cells to produce chemokines like IL-8, which increases infiltration of neutrophils and monocytes (Bowdish et al. 2005). Limulus polyphemus anti-lipopolysaccharide factor (LALF) derived peptide also have immmunomodulatory properties. They induced the cells to produce immunomodulating cytokines (Vallespi et al. 2000). The insect-derived peptide CEMA suppressed LPS-induced inflammatory gene expression in Murine RAW macrophages (IL-1ß, NO) (Scott and Hancock 2000). LALF and tachyplesin also show immmunomodulatory properties (Tanaka et al. 1982; Nakamura et al. 1988). Several analogs of Bac2a were designed by QSAR approach and their immunomodulatory properties were checked. For example, IDR-1(KSRIVPAIPVSLL-NH2) derived from Bac2a, have the ability to enhance chemokine induction and suppress LPS-stimulated pro-inflammatory cytokines (Haney and Hancock 2013). Another peptide, IDR-1002 (VQRWLIVWRIRK-NH2) is a much stronger inducer of chemokine production (Nijnik et al. 2010). Moreover, IDR-1018 (VRLIVAVRIWRR-NH2), also show good immunomodulatory properties (Haney and Hancock 2013).

CHAPTER-5
MODES OF ACTION OF ANTIMICROBIAL PEPTIDES

5. MODES OF ACTION OF ANTIMICROBIAL PEPTIDES

Several parameters such as charge, hydrophobicity, composition of amino acids, composition of membrane lipids, and peptide concentration affects the biological activities of peptides. Several experiments have been performed on model membranes and based on these experiments several models have been proposed. Until now broadly two type of mechanism have been proposed (1) Membrane-disruptive (2) Non-membrane-disruptive (Intracellular targets). Bacterial membranes are negatively charged because of the presence of phosphatidylglycerol. On the other hand, mammalian membranes are zwitterionic and neutral in charge. Antimicrobial peptides are amphipathic and cationic in nature and because they interact differently with model membranes of bacterial and mammalian cells. In membrane disruption mechanism, AMPs cause cell lysis by rupturing the membrane by utilizing several mechanisms such as Carpet mechanism, Barrel-stave mechanism and Toroidal or Wormhole mechanism and Aggregate models. In non-disruption model, AMPs translocate into cells without damaging the membrane and interact with intracellular components to inhibit bacterial growth (Also known as alternative mechanism of action of AMPs). We will discuss briefly about the several types of mechanism, which have been proposed till date.

5.1 Carpet Mechanism

Some AMPs bind to the membrane through electrostatic interaction and spread onto the lipid bilayer surface in a carpet like fashion. When a sufficient quantity of peptide concentration is reached they formed transient pores or dissolve the membrane like detergents (Fig.5.1.1). Melittins, cecropins, ovisprin and LL-37 are few examples of AMPs, which follow carpet mechanism to cause cell lysis

CHAPTER-5: MODES OF ACTION OF ANTIMICROBIAL PEPTIDES

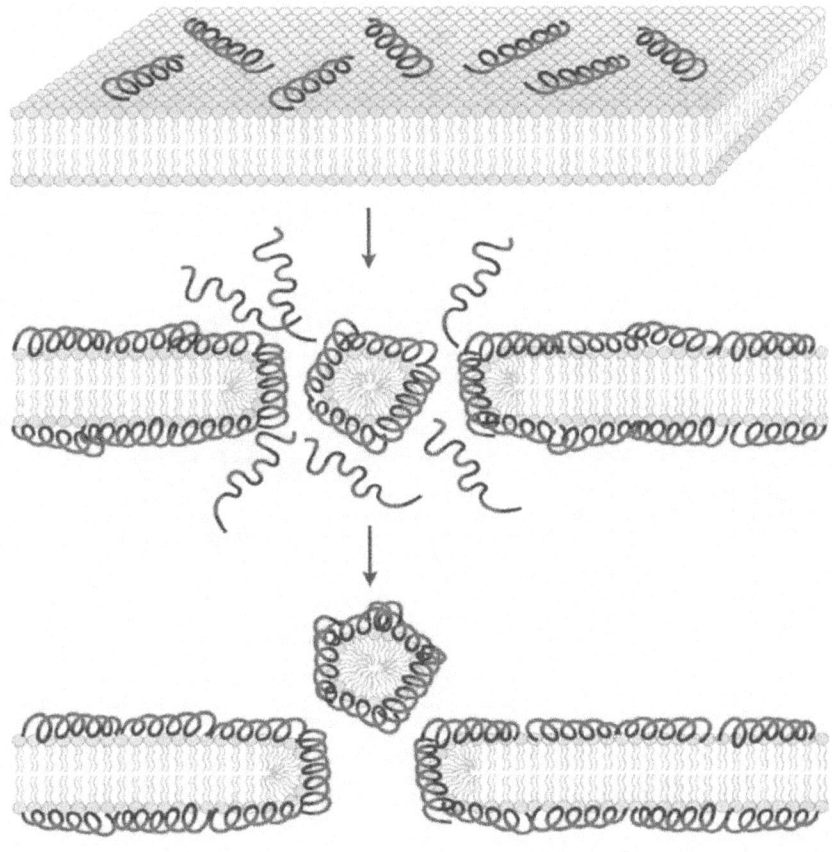

Fig.-5.1.1: The carpet model of antimicrobial-induced killing. Hydrophilic and hydrophobic regions are shown red and blue in color respectively. This figure adopted from (Brogden 2005). See text for details.

CHAPTER-5: MODES OF ACTION OF ANTIMICROBIAL PEPTIDES

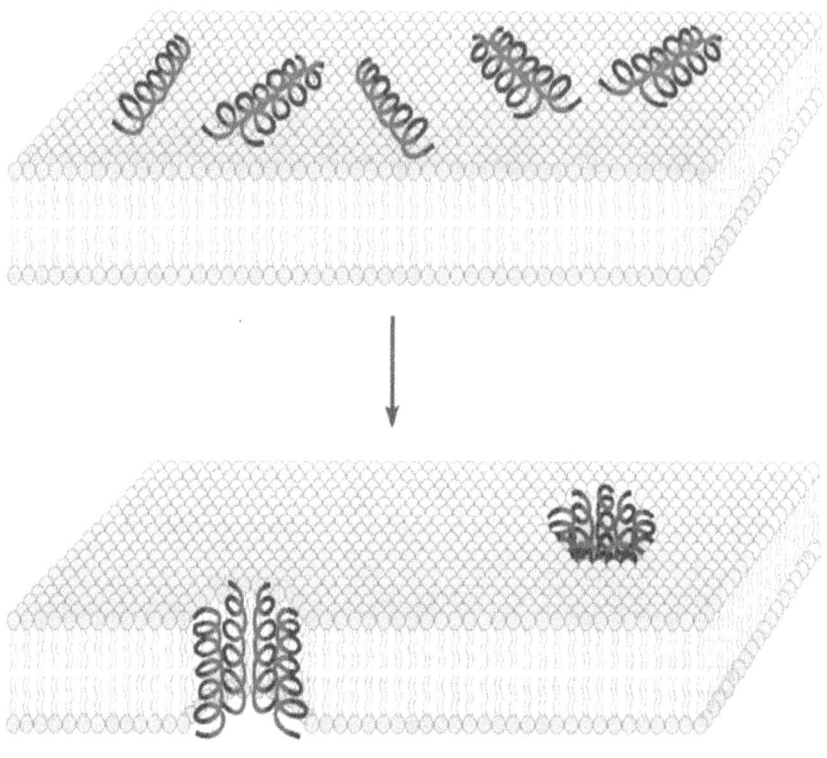

Figure-5.2.1: The Barrel-Stave model of antimicrobial-induced killing. Hydrophilic and hydrophobic regions are shown red and blue in color respectively. This figure adopted from (Brogden 2005). See text for details.

CHAPTER-5:MODES OF ACTION OF ANTIMICROBIAL PEPTIDES

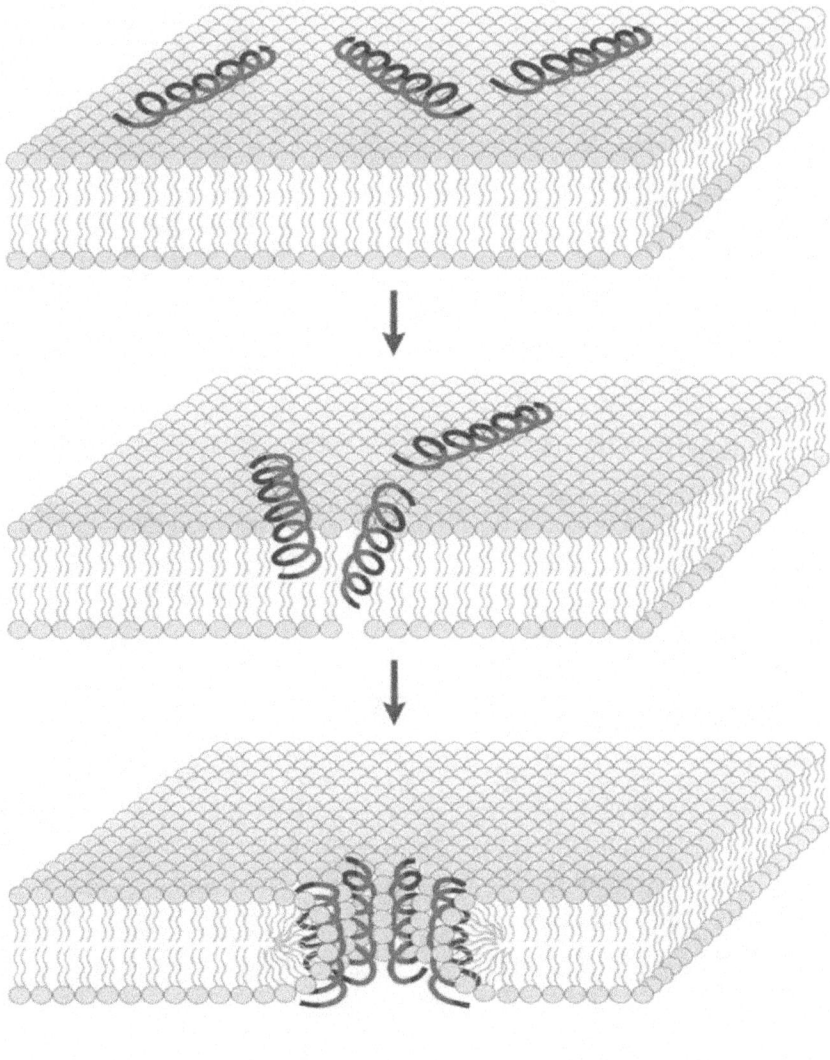

Fig.-5.3.1:The toroidal model of antimicrobial-induced killing. Hydrophilic and hydrophobic regions are shown red and blue in color respectively. This figure adopted from (Brogden 2005). See text for details.

CHAPTER-5: MODES OF ACTION OF ANTIMICROBIAL PEPTIDES

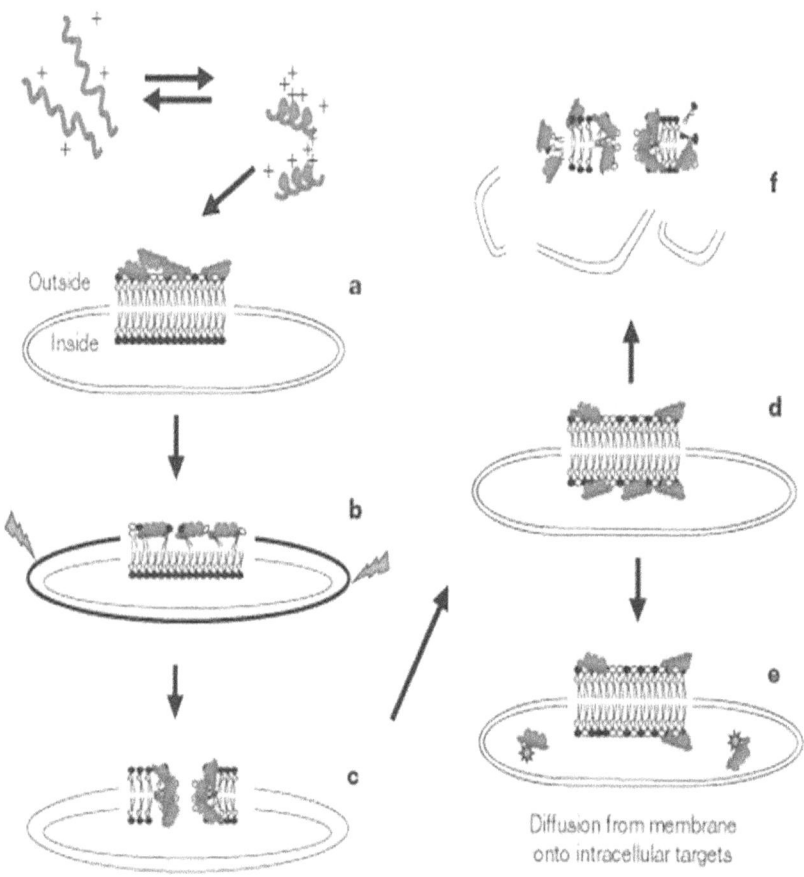

Fig.-5.5.1: The Shai-Matsuzaki-Huang model for modes of action of an antimicrobial peptide. This figure adopted from (Zasloff 2002). See text for details.

(Shai and Oren 2001; Oren and Shai 1998; Shai 1999, 2002; Papo and Shai 2003a; Gazit et al. 1995; Yeaman and Yount 2003).

CHAPTER-5: MODES OF ACTION OF ANTIMICROBIAL PEPTIDES

5.2 Barrel-Stave Mechanism

This is a well-known model to understand the mode of action of some

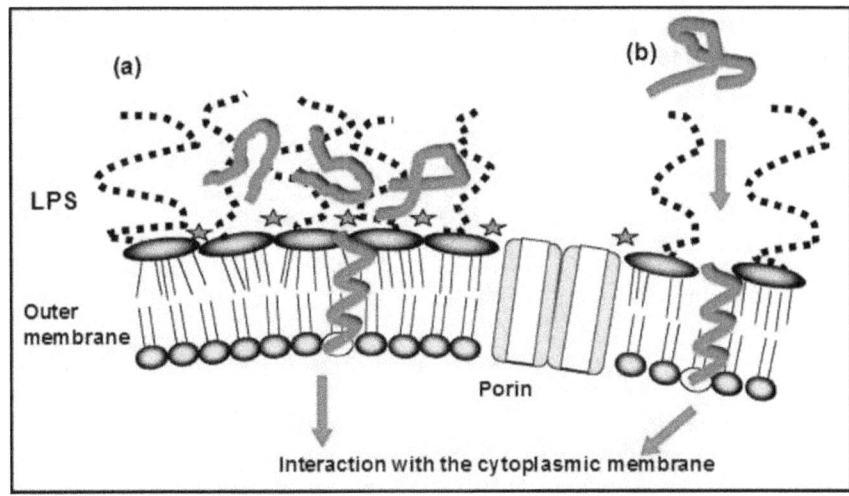

Fig.-5.6.1: Self-promoted uptake of cationic peptides model for an antimicrobial peptide. This figure adopted from (Hancock 2001). See text for details.

AMPs. According to this model, peptide first assemble on the surface of lipid bilayer and then afterward form the pores in the membrane (Fig.5.2.1). Some peptides such as pardaxin and alamethicin utilize this mechanism to kill the microorganism (Rapaport and Shai 1991; Shai et al. 1990; Rizzo et al. 1987).

5.3 Toroidal Pore or Wormhole Mechanism

This is also one of the well-known models to elucidate the mode of action of AMPs. Several peptides such as magainin, melittin and LL-37 use this mechanism to kill microorganism (Yeaman and Yount 2003). In this model, polar part of the peptide is always associated with polar part of lipid when they are perpendicularly inserted into the lipid bilayer (Fig.5.3.1). In this model, peptide and lipid both bend continuously from top to bottom in a toroidal shape and in this process hydrophilic part of both peptide and lipid face towards pore (Jenssen et al. 2006). This model has been studied extensively in some review article (Brogden

2005).

5.4 Aggregate Model
This model has been proposed by Hancock and his lab members to explain the mode of action of certain peptide and it is very much clear

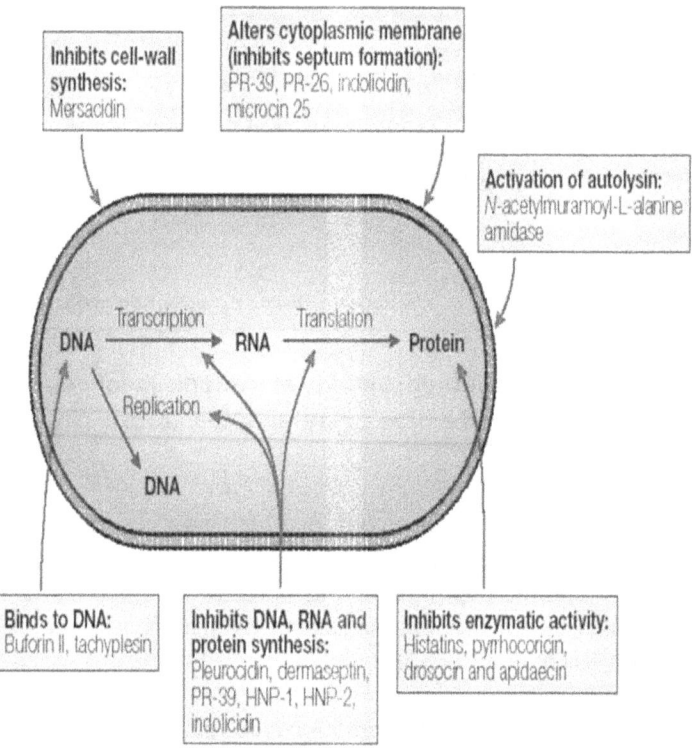

Intracellular killing mechanisms of AMPs

(*Nat. Rev. Micro.* (2005) 3: 238-250)

Fig.5.7.1: Alternative mechanism of action of an antimicrobial peptide. This figure adopted from (Brogden 2005). See text for details.

that this model has some similarity with toroidal model. According to

CHAPTER-5:MODES OF ACTION OF ANTIMICROBIAL PEPTIDES

this model, peptide reorient to span the membrane as an aggregate, and this aggregate is a mixture of lipids and peptide, which is a type of micelle. In this mechanism, peptide does not adopt a particular orientation (Powers et al. 2005). For example horse shoe crab peptide, and polyphemusin utilize this mechanism to kill bacteria (Jenssen et al. 2006).

5.5 Shai-Matsuzaki-Huang (SMH) model
According to this model, cationic peptides interact with the outer membrane of bacterial cell because of the presence of negatively charged lipids, this is followed by the collapse of membrane, and physical disruption of cell membrane and in some cases peptide enters into cells to target cellular components (Fig.5.5.1). Certain plant defensin and nisin use this mechanism to kill bacteria. For detail, please read the review article published by (Zasloff 2002).

5.6 Self-promoted Uptake of Cationic Peptides
Numerous models have been suggested that how the peptides kill the bacteria (Hancock and Chapple 1999; Huang 2000). Among these, Self-promoted uptake of cationic peptides is one of the models proposed by Hancock group to explain gram-negative bacteria killing (Fig.5.6.1). It is very well establish that gram-negative bacteria contain additional outer membrane barrier. In this model, cationic peptide interact with bacterial membrane because of presence of the negative charge on outer surface due to the presence of anionic lipopolysaccharide. The interaction of cationic peptides with outer membrane neutralizes the surface, which causes cracks in outer membrane through which peptide can cross. Otherwise, they can also bind to a divalent cationic binding site on lipopolysaccharides and destabilize the membrane. After crossing the outer membrane, they interact with the negatively charged surface of the cytoplasmic membrane and insert inside the cell (Hancock 2001).

5.7 Alternative Mechanism of Action
We know that conventional antibiotic kill the bacteria by specifically blocking cell wall synthesis, inhibiting protein synthesis, inhibiting nucleic acid synthesis, inhibiting metabolic pathway and interference with cell membrane integrity (Fig.5.7.1). Therefore, chances of the development of bacterial resistance are easier in case of conventional antibiotic. On the other hand, most of antimicrobial peptides kill the

CHAPTER-5: MODES OF ACTION OF ANTIMICROBIAL PEPTIDES

bacteria by disrupting the whole integrity of their cell and hence leave very little scope for development of bacterial resistance. Some of antimicrobial peptides kill the bacteria similar to conventional antibiotic. These peptides translocate across the cell membrane, accumulate intracellularly, and kill the bacteria by various intracellular process, including inhibition of nucleic acid synthesis, protein synthesis, enzymatic activity, and cell wall synthesis (Brogden 2005). Several peptides such as buforin II, derivatives of pleurocidin, dermaseptin, human defensin, HNP-1 and indolicidin kill bacteria by inhibiting nucleic acid synthesis (Friedrich et al. 2001, Park et al., 1998, Subbalakshmi&. Sitaram, 1998, Patrzykat et al., 2002, Lehrer et al., 1989). Moreover, Pleurocidin, dermaseptin and indolicidin kill bacteria by inhibiting protein synthesis. Pyrrhocidin, histatins, drosocin, and apidaecin kill bacteria by Inhibition of cellular enzymatic activity (Brogden, 2005, Otvos et al., 2000, Kragol et al., 2001). Mersacidin inhibit cell wall synthesis. PR-39, PR-26, and microcin kill bacteria by altering cytoplasmic membrane (Brogden 2005).

CHAPTER-6
APPLICATION OF ANTIMICROBIAL PEPTIDES

6. APPLICATION OF ANTIMICROBIAL PEPTIDES

A plenty of antimicrobial peptides have been isolated from different organism, including plants, animals, bacteria, and protozoa. These antimicrobial peptides show diverse function. They are also known as primary defense of our body. Our body also produces a variety of antimicrobial peptides, which protects our body from bacterial infection. Antimicrobial peptides show a wide range of application, some of which are described below in brief.

Table-6.1.1: Antimicrobial peptides in clinical trials (Fox 2013).

Peptide Name	Company	Stage of development	Medical use
Magainin peptide/ pexiganan acetate	Dipexium Pharma (White Plains, New York)/MacroChem/Genaera	3	Diabetic foot ulcers
Omiganan	BioWest Therapeutics/Maruho (Vancouver)	2	Rosacea
OP-145	OctoPlus (Leiden, The Netherlands)	2	Chronic bacterial middle-ear infection

CHAPTER-6: APPLICATION OF ANTIMICROBIAL PEPTIDES

Novexatin	NovaBiotics (Aberdeen, UK)	1/2	Fungal infections of the toe, and nail
Lytixar (LTX-109)	Lytix Biopharma (Oslo)	1/2	Nasally colonized MRSA
NVB302	Novacta (Welwyn Garden City, UK)	1	*C. difficile*
MU1140	Oragenics (Tampa, Florida)	Preclinical	Gram-positive bacteria (MRSA, *C. difficile*)
Arenicin	Adenium Biotech Copenhagen	Preclinical	Multiresistant Gram-positive bacteria
Avidocin and purocin	AvidBiotics (S. San Francisco, California)	Preclinical	Narrow spectrum antibiotic for human health and food safety
IMX924	Iminex (Coquitlam, British Columbia,	Preclinical	Gram-negative and

CHAPTER-6:APPLICATION OF ANTIMICROBIAL PEPTIDES

	Canada)		Gram-positive bacteria (improves survival and reduces tissue damage)

6.1 Therapeutic uses of Antimicrobial Peptides

There is a great interest in antimicrobial peptides from the last few years because of their unique mode of actions. It is well documented that these antimicrobial peptides have broad spectrum activities against gram-negative and gram-positive bacteria. These are also recognized as ancient weapons to fight against a wide range of bacterial infections. Moreover, increasing bacterial resistance against conventional antibiotic has motivated the pharmaceutical industry to develop novel therapeutic agents. The unique properties of antimicrobial peptides could make them a better alternative to conventional antibiotic. Several efforts have been made to develop peptide-based antibiotic. Several antimicrobial peptides are in clinical and preclinical stages. These peptides are summarized in table-6.1.1.

Table-6.2.1: Examples of selected synthetic antimicrobial peptides, which are active against plant pathogens (Montesinos 2007).

Peptide name	Source	Sequence
Pexiganan	Magainin	GIGKFLKKAKKFGKAFVKILKK-NH2
Iseganan	Protegrin	RGGLCYCRGRFCVCVGR-NH2
D32R	Thionin	KSCCRNTWARNCYNVCRLPGTISREI

CHAPTER-6: APPLICATION OF ANTIMICROBIAL PEPTIDES

		CAKKCRCKIISGTTCPSDYPK
Pen4-1	Penaedin	HSSGYTRPLRKPSRPIFIRPIGCDVCYGI PSSTARLCCFRYGDCHL-NH2
MB-39	Cecropin	HQPKWKVFKKIEVVGRNIRNGI VKAGPAIAVLGEAKALG
CAMEL	Cecropin–melittin hybrid	KWKLFKKIGAVLKVL-NH2
TPY	Tachyplesin	KWVFRVNYRGIKYRRQR

6.2 Application of Antimicrobial Peptides in Agriculture

Bacteria, viruses and fungal infections cause most of plant diseases. These microorganisms are responsible for remarkable decrease or losses of crop production. The plant diseases also decrease quality or safety of agricultural products (Meng et al. 2010). The pesticides, which are available in the market, are mainly used for controlling these microorganisms. Because of high toxicity and increase negative impact on the environment, scientist and researcher are searching an alternative to overcome this problem. The antimicrobial peptides can be one of the therapeutic agent to replace pesticide because of their broad-spectrum activities. Several antimicrobial peptides such as Bacteriocins, fungal defensins and cyclopeptides have the capacity to kill the phytopathogen. Some pseudopeptides, and synthetic AMPs have been synthesized which are highly active against the plant pathogens (Montesinos 2007). A few synthetic AMPs, which are active against plant pathogen, are summarized in table-6.2.1.

Nowadays, several strategies have been established to develop disease-resistant plants. Transformation with genes encoding AMPs in plants is a promising approach to develop resistance against phytopathogens (Meng et al. 2010). In recent years, several antimicrobial peptide genes have been successfully expressed in plants to protect against pathogen such as Cecropins A and B, magainin, tachyplesin, seliomicin, drosomycin, sarcotoxin, and plant defensins. Some synthetic AMPs gene have been also introduced in plants for phytopathogen

CHAPTER-6: APPLICATION OF ANTIMICROBIAL PEPTIDES

protection (Table-6.2.2), including Myp30, SB-37, Rev4, MB-39 and MsrA1 (Montesinos 2007; Meng et al. 2010).

Table-6.2.2: Selected examples of antimicrobial peptides expressed in transgenic plants to protect against phytopathogen (Montesinos 2007; Meng et al. 2010).

Antimicrobial peptide	source	Plant transformed
Cecropin A, B	Moth haemolynph	Rice
Tachyplesin	Crab haemolynph	Potato
Drosomycin	Insect defensin	Tobacco
Sarcotoxin IA	Fruit fly haemolynph	Tobacco
Magainin	Frog skin	Tobacco
Alf-AFP (Plant defensing)	Alfalfa defensin	Potato
Myp30	Magainin analogue	Tobacco
SB-37	Cecropin analogue	Potato, apple
Rev4	Indolicidin analogue	Tobacco/ Arabidopsis
MB-39	Cecropin analogue	Apple
MsrA1	Cecropin–melittin hybrid	Potato

6.3 Application of Antimicrobial Peptides in Food Industry

It is very well known facts that chemical is widely used for a long time as a food preservative in the food industry. Some of the well known food preservative chemicals are benzene-carboxylic acid, sorbic acid, and parahydroxybenzoate ester (Meng et al. 2010). The chemical food

CHAPTER-6: APPLICATION OF ANTIMICROBIAL PEPTIDES

preservative may be not be appropriate for human health and they can alter food taste. Therefore, the food industry is looking to replace these chemical food preservative with some natural resources which can be good for health and maintain the taste of food. Because of low toxicity, safe for human and having broad-spectrum activities, AMPs can be used as a potential food preservative. Several antimicrobial peptides are used for food preservatives, which are isolated from different organisms. Bacteriocins are antimicrobial peptides, isolated from bacteria. Nisin, is a type of Bacteriocins which are widely used for dairy and the canning industries as a biopresertive (Meng et al. 2010). Other Bacteriocins such as lactococcins, pediocins, lactacins, and leucocinA are used in food industry as a preservative (Meng et al. 2010).

Table-6.6.1: Selected examples of antimicrobial peptides used as a drug delivery vector (Pushpanathan et al. 2013; Doehlemann et al. 2009).

Antimicrobial peptide	Source	Sequence
LL-37	Human	LLGDFFRKSKEKIGKEFKRIVQRIKDF LRNLVPRTES
Magainin 2	Frog	GIGKFLHSAKKFGKAFVGEIMNS
Buforin 2	Toad	TRSSRAGLQFPVGRVHRLLRK
Pyrrhocoricin	European fire bug	VDKGSYLPRPTPPRPIYNRN
Bac7	Bovine neutrophils	RRIRPRPPRLPRPRPRPLPFPRPGPR PIPRPLPFPRPGPRPIPRPLPFPRPGP RPIPRPL
Penetratin	Drosophila	RQIKIWFQNRRMKWKKGG
Pep1	Vacuum-packed sausages	KETWWETWWTEWSQPKKKRKV

CHAPTER-6: APPLICATION OF ANTIMICROBIAL PEPTIDES

| MMGP1 | Marine metagenome | MLWSASMRIFASAFSTRGLGTRML MYCSLPSRCWRK |

6.4 Antimicrobial Peptide in Surface Coating

Antimicrobial coatings are applied to counters, walls, door handles, and other high-touch areas to inhibit the growth of microorganism. In hospitals, these coatings are applied to biomedical devices to prevent bacterial growth. It is well accepted that antimicrobial peptide coating reduces bacterial colonization and biofilm formation. For example, hLF1-11 immobilized onto titanium possess excellent anti-biofilm properties (Wang et al. 2015). Another, antimicrobial peptides temporin-SHf show broad-spectrum activity against microorganism (Wang et al. 2015). Melimine coating on eye lenses show antibacterial properties (Wang et al. 2015; Cole et al. 2010). Moreover, another peptide SESB2V immobilized on titanium surfaces prevents bacterial infection (Tan et al. 2012).

6.5 Application of Antimicrobial peptide in Biosensors and Detection

The majority of rapid detection systems uses antibodies or nucleic acid probes for sensor and detection. However, antibodies or nucleic acid probes based sensor has some limitation. Antimicrobial peptides can be also used as a biosensors and diagnostic agents. Several antimicrobial peptides such as maganin1, dermaseptin and class IIa bacteriocins (Mannoor et al. 2010; Zampa et al. 2012; Wang et al. 2015) are used as a biosensor.

6.6 Antimicrobial Peptides in Drug Delivery System

Majority of antimicrobial peptides interact with the cell membrane of the bacteria cells and formed pores therein. A few antimicrobial peptides are non-lytic cell penetrating peptides, which could enter the cells without causing damage to the cell membranes and therefore, considered as a potential candidates for drug delivery vectors. Several antimicrobial peptides have been successfully used as a drug delivery vector (Table-6.6.1) including LL-37, TP10, and pVEC, magainin 2, buforin 2, SynB, pyrrhocoricin and Bac7 (Pushpanathan et al. 2013; Doehlemann et al. 2009).

REFERENCES

Agerberth B, Gunne H, Odeberg J, Kogner P, Boman HG, Gudmundsson GH (1995) FALL-39, a putative human peptide antibiotic, is cysteine-free and expressed in bone marrow and testis. Proceedings of the National Academy of Sciences 92 (1):195-199

Ahmad A, Ahmad E, Rabbani G, Haque S, Arshad M, Hasan Khan R (2012) Identification and design of antimicrobial peptides for therapeutic applications. Current Protein and Peptide Science 13 (3):211-223

Ahmad A, Asthana N, Azmi S, Srivastava RM, Pandey BK, Yadav V, Ghosh JK (2009a) Structure–function study of cathelicidin-derived bovine antimicrobial peptide BMAP-28: design of its cell-selective analogs by amino acid substitutions in the heptad repeat sequences. Biochimica et Biophysica Acta (BBA)-Biomembranes 1788 (11):2411-2420

Ahmad A, Azmi S, Ghosh JK (2011) Studies on the assembly of a leucine zipper antibacterial peptide and its analogs onto mammalian cells and bacteria. Amino acids 40 (2):749-759

Ahmad A, Azmi S, Srivastava RM, Srivastava S, Pandey BK, Saxena R, Bajpai VK, Ghosh JK (2009b) Design of nontoxic analogues of cathelicidin-derived bovine antimicrobial peptide BMAP-27: the role of leucine as well as phenylalanine zipper sequences in determining its toxicity. Biochemistry 48 (46):10905-10917

Ahmad A, Yadav SP, Asthana N, Mitra K, Srivastava SP, Ghosh JK (2006) Utilization of an amphipathic leucine zipper sequence to design antibacterial peptides with simultaneous modulation of toxic activity against human red blood cells. Journal of Biological Chemistry 281 (31):22029-22038

Akuffo H, Hultmark D, Engstöm A, Frohlich D, Kimbrell D (1998) Drosophila antibacterial protein, cecropin A, differentially affects non-bacterial organisms such as Leishmania in a manner different from other amphipathic peptides. International journal of molecular medicine 1:77-82

Andrews JM (2001) Determination of minimum inhibitory concentrations. Journal of antimicrobial Chemotherapy 48 (suppl 1):5-16

Asaria P, MacMahon E (2006) Measles in the United Kingdom: can we eradicate it by. BMJ 333:890-895

Avrahami D, Shai Y (2003) Bestowing antifungal and antibacterial activities by lipophilic acid conjugation to d, l-amino acid-containing antimicrobial peptides: a plausible mode of action. Biochemistry 42 (50):14946-14956

BABASAKI K, TAKAO T, SHIMONISHI Y, KURAHASHI K (1985)

REFERENCES

Subtilosin A, a new antibiotic peptide produced by Bacillus subtilis 168: isolation, structural analysis, and biogenesis. Journal of biochemistry 98 (3):585-603

Barbosa Pelegrini P, del Sarto RP, Silva ON, Franco OL, Grossi-de-Sa MF (2011) Antibacterial peptides from plants: what they are and how they probably work. Biochemistry Research International 2011

Bennallack PR, Burt SR, Heder MJ, Robison RA, Griffitts JS (2014) Characterization of a novel plasmid-borne thiopeptide gene cluster in Staphylococcus epidermidis strain 115. Journal of bacteriology 196 (24):4344-4350

Bera A, Singh S, Nagaraj R, Vaidya T (2003) Induction of autophagic cell death in Leishmania donovani by antimicrobial peptides. Molecular and biochemical parasitology 127 (1):23-35

Bhunia A, Domadia PN, Torres J, Hallock KJ, Ramamoorthy A, Bhattacharjya S (2010) NMR structure of pardaxin, a pore-forming antimicrobial peptide, in lipopolysaccharide micelles mechanism of outer membrane permeabilization. Journal of biological chemistry 285 (6):3883-3895

Blondelle SE, Takahashi E, Dinh KT, Houghten R (1995) The antimicrobial activity of hexapeptides derived from synthetic combinatorial libraries. Journal of applied bacteriology 78 (1):39-46

Boman H (2003) Antibacterial peptides: basic facts and emerging concepts. Journal of internal medicine 254 (3):197-215

Boman HG, Agerberth B, Boman A (1993) Mechanisms of action on Escherichia coli of cecropin P1 and PR-39, two antibacterial peptides from pig intestine. Infection and Immunity 61 (7):2978-2984

Bowdish D, Davidson D, Hancock R (2006) Immunomodulatory properties of defensins and cathelicidins. In: Antimicrobial Peptides and Human Disease. Springer, pp 27-66

Bowdish DM, Davidson DJ, Scott MG, Hancock RE (2005) Immunomodulatory activities of small host defense peptides. Antimicrobial agents and chemotherapy 49 (5):1727-1732

Brand GD, Leite JRS, Silva LP, Albuquerque S, Prates MV, Azevedo RB, Carregaro V, Silva JS, Sá VC, Brandão RA (2002) Dermaseptins from Phyllomedusa oreades andphyllomedusa distincta anti-trypanosoma cruzi activity without cytotoxicity to mammalian cells. Journal of Biological Chemistry 277 (51):49332-49340

Broekaert WF, Cammue BP, De Bolle MF, Thevissen K, De Samblanx GW, Osborn RW, Nielson K (1997) Antimicrobial peptides from plants. Critical reviews in plant sciences 16 (3):297-323

REFERENCES

Brogden KA (2005) Antimicrobial peptides: pore formers or metabolic inhibitors in bacteria? Nature Reviews Microbiology 3 (3):238-250

Bulet P, Stocklin R (2005) Insect antimicrobial peptides: structures, properties and gene regulation. Protein and peptide letters 12 (1):3-11

Bulet P, Stöcklin R, Menin L (2004) Anti-microbial peptides: from invertebrates to vertebrates. Immunological reviews 198 (1):169-184

Bulet P, Urge L, Ohresser S, Hetru C, Otvos L (1996) Enlarged Scale Chemical Synthesis and Range of Activity of Drosocin, an O-Glycosylated Antibacterial Peptide of Drosophila. European journal of biochemistry 238 (1):64-69

Capparelli R, Amoroso MG, Palumbo D, Iannaccone M, Faleri C, Cresti M (2005) Two plant puroindolines colocalize in wheat seed and in vitro synergistically fight against pathogens. Plant molecular biology 58 (6):857-867

Čeřovský V, Hovorka O, Cvačka J, Voburka Z, Bednárová L, Borovičková L, Slaninová J, Fučík V (2008) Melectin: a novel antimicrobial peptide from the venom of the cleptoparasitic bee Melecta albifrons. ChemBioChem 9 (17):2815-2821

Chan DI, Prenner EJ, Vogel HJ (2006) Tryptophan- and arginine-rich antimicrobial peptides: structures and mechanisms of action. Biochimica et biophysica acta 1758 (9):1184-1202. doi:10.1016/j.bbamem.2006.04.006

Chapman T, Kinsman O, Houston J (1992) Chitin biosynthesis in Candida albicans grown in vitro and in vivo and its inhibition by nikkomycin Z. Antimicrobial agents and chemotherapy 36 (9):1909-1914

Charnet P, Molle G, Marion D, Rousset M, Lullien-Pellerin V (2003) Puroindolines form ion channels in biological membranes. Biophysical journal 84 (4):2416-2426

Chen J-J, Chen G-H, Hsu H-C, Li S-S, Chen C-S (2004) Cloning and functional expression of a mungbean defensin VrD1 in Pichia pastoris. Journal of agricultural and food chemistry 52 (8):2256-2261

Chinchar V, Bryan L, Silphadaung U, Noga E, Wade D, Rollins-Smith L (2004) Inactivation of viruses infecting ectothermic animals by amphibian and piscine antimicrobial peptides. Virology 323 (2):268-275

Clara A, Manjramkar DD, Reddy VK (2004) Preclinical evaluation of magainin-A as a contraceptive antimicrobial agent. Fertility and

REFERENCES

sterility 81 (5):1357-1365

Cociancich S, Bulet P, Hetru C, Hoffmann JA (1994a) The inducible antibacterial peptides of insects. Parasitology today (Personal ed) 10 (4):132-139

Cociancich S, Dupont A, Hegy G, Lanot R, Holder F, Hetru C, Hoffmann JA, Bulet P (1994b) Novel inducible antibacterial peptides from a hemipteran insect, the sap-sucking bug Pyrrhocoris apterus. The Biochemical journal 300 (Pt 2):567-575

Cole AM, Waring AJ (2002) The role of defensins in lung biology and therapy. American Journal of Respiratory Medicine 1 (4):249-259

Cole N, Hume EB, Vijay AK, Sankaridurg P, Kumar N, Willcox MD (2010) In vivo performance of melimine as an antimicrobial coating for contact lenses in models of CLARE and CLPU. Investigative ophthalmology & visual science 51 (1):390-395

Conde R, Zamudio FZ, Rodríguez MH, Possani LD (2000) Scorpine, an anti-malaria and anti-bacterial agent purified from scorpion venom. FEBS letters 471 (2-3):165-168

Conlan BF, Gillon AD, Barbeta BL, Anderson MA (2011) Subcellular targeting and biosynthesis of cyclotides in plant cells. American journal of botany 98 (12):2018-2026

Cotter PD, Hill C, Ross RP (2005a) Bacterial lantibiotics: strategies to improve therapeutic potential. Current Protein and Peptide Science 6 (1):61-75

Cotter PD, Hill C, Ross RP (2005b) Bacteriocins: developing innate immunity for food. Nature Reviews Microbiology 3 (10):777-788

Craik DJ (2012) Host-defense activities of cyclotides. Toxins 4 (2):139-156

Craik DJ, Daly NL, Bond T, Waine C (1999) Plant cyclotides: a unique family of cyclic and knotted proteins that defines the cyclic cystine knot structural motif. Journal of molecular biology 294 (5):1327-1336

Cruciani RA, Barker JL, Zasloff M, Chen HC, Colamonici O (1991) Antibiotic magainins exert cytolytic activity against transformed cell lines through channel formation. Proceedings of the National Academy of Sciences 88 (9):3792-3796

Daly NL, Clark RJ, Plan MR, Craik DJ (2006) Kalata B8, a novel antiviral circular protein, exhibits conformational flexibility in the cystine knot motif. Biochemical Journal 393 (3):619-626

Daniels R, Nicoll LH (2011) Contemporary medical-surgical nursing. Cengage Learning,

Date-Ito A, Kasahara K, Sawai H, Chigusa SI (2002) Rapid evolution of the male-specific antibacterial protein andropin gene in Drosophila.

REFERENCES

Journal of molecular evolution 54 (5):665-670

Dawson RM, Liu CQ (2009) Cathelicidin peptide SMAP-29: comprehensive review of its properties and potential as a novel class of antibiotics. Drug Development Research 70 (7):481-498

De Lucca A, Bland J, Jacks T, Grimm C, Walsh T (1998) Fungicidal and binding properties of the natural peptides cecropin B and dermaseptin. Medical mycology 36 (5):291-298

De Lucca AJ, Walsh TJ (1999) Antifungal peptides: novel therapeutic compounds against emerging pathogens. Antimicrobial agents and chemotherapy 43 (1):1-11

De Smet K, Contreras R (2005) Human antimicrobial peptides: defensins, cathelicidins and histatins. Biotechnology letters 27 (18):1337-1347

Debono M, Gordee RS (1994) Antibiotics that inhibit fungal cell wall development. Annual Reviews in Microbiology 48 (1):471-497

DeLucca AJ, Bland JM, Jacks TJ, Grimm C, Cleveland TE, Walsh TJ (1997) Fungicidal activity of cecropin A. Antimicrobial agents and chemotherapy 41 (2):481-483

Destoumieux D, Munoz M, Bulet P, Bachere E (2000) Penaeidins, a family of antimicrobial peptides from penaeid shrimp (Crustacea, Decapoda). Cellular and Molecular Life Sciences CMLS 57 (8-9):1260-1271

Dhama K, Saminathan M, Jacob SS, Singh M, Karthik K, Amarpal, Tiwari R, Sunkara LT, Malik YS, Singh RK (2015) Effect of immunomodulation and immunomodulatory agents on health with some bioactive principles, modes of action and potent biomedical applications. INTERNATIONAL JOURNAL OF PHARMACOLOGY 11 (4):253-290

Doehlemann G, Van Der Linde K, Aßmann D, Schwammbach D, Hof A, Mohanty A, Jackson D, Kahmann R (2009) Pep1, a secreted effector protein of Ustilago maydis, is required for successful invasion of plant cells. PLoS Pathog 5 (2):e1000290

Dubos RJ (1939) Studies on a bactericidal agent extracted from a soil bacillus: I. Preparation of the agent. Its activity in vitro. The Journal of experimental medicine 70 (1):1

Duquesne S, Destoumieux-Garzón D, Peduzzi J, Rebuffat S (2007) Microcins, gene-encoded antibacterial peptides from enterobacteria. Natural product reports 24 (4):708-734

Ellerby HM, Arap W, Ellerby LM, Kain R, Andrusiak R, Del Rio G, Krajewski S, Lombardo CR, Rao R, Ruoslahti E (1999) Anti-cancer activity of targeted pro-apoptotic peptides. Nature medicine 5 (9):1032-1038

REFERENCES

Falla TJ, Karunaratne DN, Hancock RE (1996) Mode of action of the antimicrobial peptide indolicidin. Journal of Biological Chemistry 271 (32):19298-19303

Feder R, Dagan A, Mor A (2000) Structure-activity relationship study of antimicrobial dermaseptin S4 showing the consequences of peptide oligomerization on selective cytotoxicity. Journal of Biological Chemistry 275 (6):4230-4238

Fehlbaum P, Bulet P, Michaut L, Lagueux M, Broekaert WF, Hetru C, Hoffmann JA (1994) Insect immunity. Septic injury of Drosophila induces the synthesis of a potent antifungal peptide with sequence homology to plant antifungal peptides. Journal of Biological Chemistry 269 (52):33159-33163

Fennell JF, Shipman WH, Cole LJ (1967) Antibacterial action of a bee venom fraction (melittin) against a penicillin-resistant staphylococcus and other microorganisms. DTIC Document,

Finlay BB, Hancock RE (2004) Can innate immunity be enhanced to treat microbial infections? Nature Reviews Microbiology 2 (6):497-504

Fox JL (2013) Antimicrobial peptides stage a comeback. Nature biotechnology 31 (5):379-382

Friedrich CL, Rozek A, Patrzykat A, Hancock RE (2001) Structure and mechanism of action of an indolicidin peptide derivative with improved activity against gram-positive bacteria. Journal of biological chemistry 276 (26):24015-24022

Fujikawa K, Suketa Y, Hayashi K, Suzuki T (1965) Chemical structure of circulin A. Cellular and Molecular Life Sciences 21 (6):307-308

Gallo RL, Kim KJ, Bernfield M, Kozak CA, Zanetti M, Merluzzi L, Gennaro R (1997) Identification of CRAMP, a cathelin-related antimicrobial peptide expressed in the embryonic and adult mouse. Journal of Biological Chemistry 272 (20):13088-13093

Gallo RL, Murakami M, Ohtake T, Zaiou M (2002) Biology and clinical relevance of naturally occurring antimicrobial peptides. Journal of Allergy and Clinical Immunology 110 (6):823-831

Ganz T, Lehrer RI (1998) Antimicrobial peptides of vertebrates. Current opinion in immunology 10 (1):41-44

Gao B, Xu J, del Carmen Rodriguez M, Lanz-Mendoza H, Hernández-Rivas R, Du W, Zhu S (2010) Characterization of two linear cationic antimalarial peptides in the scorpion Mesobuthus eupeus. Biochimie 92 (4):350-359

Gaspar D, Veiga AS, Castanho MA (2014) From antimicrobial to anticancer peptides. A review. New edge of antibiotic development: antimicrobial peptides and corresponding resistance:24

REFERENCES

Gazit E, Boman A, Boman HG, Shai Y (1995) Interaction of the mammalian antibacterial peptide cecropin P1 with phospholipid vesicles. Biochemistry 34 (36):11479-11488

Gennaro R, Zanetti M (2000) Structural features and biological activities of the cathelicidin-derived antimicrobial peptides. Peptide Science 55 (1):31-49

Gordon YJ, Romanowski EG, McDermott AM (2005) A review of antimicrobial peptides and their therapeutic potential as anti-infective drugs. Current eye research 30 (7):505-515

Gueguen Y, Bernard R, Julie F, Paulina S, Delphine D-G, Franck V, Philippe B, Evelyne B (2009) Oyster hemocytes express a proline-rich peptide displaying synergistic antimicrobial activity with a defensin. Molecular immunology 46 (4):516-522

Gustafson KR, Walton LK, Sowder RC, Johnson DG, Pannell LK, Cardellina JH, Boyd MR (2000) New Circulin Macrocyclic Polypeptides from Chassalia p arvifolia 1. Journal of natural products 63 (2):176-178

Gutsmann T, Hagge SO, Larrick JW, Seydel U, Wiese A (2001) Interaction of CAP18-derived peptides with membranes made from endotoxins or phospholipids. Biophysical Journal 80 (6):2935-2945

Haines LR, Thomas JM, Jackson AM, Eyford BA, Razavi M, Watson CN, Gowen B, Hancock RE, Pearson TW (2009) Killing of trypanosomatid parasites by a modified bovine host defense peptide, BMAP-18. PLoS Negl Trop Dis 3 (2):e373

Hale JD, Hancock RE (2007) Alternative mechanisms of action of cationic antimicrobial peptides on bacteria. Expert review of anti-infective therapy 5 (6):951-959

Hallock YF, Sowder RC, Pannell LK, Hughes CB, Johnson DG, Gulakowski R, Cardellina JH, Boyd MR (2000) Cycloviolins AD, Anti-HIV Macrocyclic Peptides from Leonia c ymosa 1. The Journal of organic chemistry 65 (1):124-128

Hancock RE (1999) Host defence (cationic) peptides: what is their future clinical potential? Drugs 57 (4):469-473

Hancock RE (2001) Cationic peptides: effectors in innate immunity and novel antimicrobials. The Lancet infectious diseases 1 (3):156-164

Hancock RE, Brown KL, Mookherjee N (2006) Host defence peptides from invertebrates—emerging antimicrobial strategies. Immunobiology 211 (4):315-322

Hancock RE, Chapple DS (1999) Peptide antibiotics. Antimicrobial agents and chemotherapy 43 (6):1317-1323

Hancock RE, Sahl H-G (2006) Antimicrobial and host-defense peptides as

REFERENCES

new anti-infective therapeutic strategies. Nature biotechnology 24 (12):1551-1557

Hancock RE, Scott MG (2000) The role of antimicrobial peptides in animal defenses. Proceedings of the national Academy of Sciences 97 (16):8856-8861

Haney EF, Hancock RE (2013) Peptide design for antimicrobial and immunomodulatory applications. Peptide Science 100 (6):572-583

Harder J, Schröder J-M (2005) Psoriatic scales: a promising source for the isolation of human skin-derived antimicrobial proteins. Journal of leukocyte biology 77 (4):476-486

Harwig SS, Kokryakov VN, Swiderek KM, Aleshina GM, Zhao C, Lehrer RI (1995) Prophenin-1, an exceptionally proline-rich antimicrobial peptide from porcine leukocytes. FEBS letters 362 (1):65-69

Hassan M, Kjos M, Nes I, Diep D, Lotfipour F (2012) Natural antimicrobial peptides from bacteria: characteristics and potential applications to fight against antibiotic resistance. Journal of applied microbiology 113 (4):723-736

Haug BE, Skar ML, Svendsen JS (2001) Bulky aromatic amino acids increase the antibacterial activity of 15-residue bovine lactoferricin derivatives. Journal of Peptide Science 7 (8):425-432

Haug BE, Svendsen JS (2001) The role of tryptophan in the antibacterial activity of a 15-residue bovine lactoferricin peptide. Journal of Peptide Science 7 (4):190-196

Helmerhorst EJ, Wim VTH, VEERMAN EC, Simoons-Smit I, Arie V (1997) Synthetic histatin analogues with broad-spectrum antimicrobial activity. Biochemical Journal 326 (1):39-45

Henderson JT, Chopko AL, Van Wassenaar PD (1992) Purification and primary structure of pediocin PA-1 produced by Pediococcus acidilactici PAC-1.0. Archives of Biochemistry and Biophysics 295 (1):5-12

Hilpert K, Volkmer-Engert R, Walter T, Hancock RE (2005) High-throughput generation of small antibacterial peptides with improved activity. Nature biotechnology 23 (8):1008-1012

Hirai Y, Yasuhara T, Yoshida H, Nakajima T, Fujino M, Kitada C (1979) A new mast cell degranulating peptide "mastoparan" in the venom of Vespula lewisii. Chemical & pharmaceutical bulletin 27 (8):1942-1944

Hiratsuka T, Mukae H, Iiboshi H, Ashitani J, Nabeshima K, Minematsu T, Chino N, Ihi T, Kohno S, Nakazato M (2003) Increased concentrations of human β-defensins in plasma and bronchoalveolar lavage fluid of patients with diffuse

REFERENCES

panbronchiolitis. Thorax 58 (5):425-430

Hori M, EGUCHI J, KAKIKI K, MISATO T (1974) STUDIES ON THE MODE OF ACTION OF POLYOXINS. VI. The Journal of antibiotics 27 (4):260-266

Hoskin DW, Ramamoorthy A (2008) Studies on anticancer activities of antimicrobial peptides. Biochimica et Biophysica Acta (BBA)-Biomembranes 1778 (2):357-375

Howell MD, Jones JF, Kisich KO, Streib JE, Gallo RL, Leung DY (2004) Selective killing of vaccinia virus by LL-37: implications for eczema vaccinatum. The Journal of Immunology 172 (3):1763-1767

Huang C, Chen H, Zierdt C (1990) Magainin analogs effective against pathogenic protozoa. Antimicrobial agents and chemotherapy 34 (9):1824-1826

Huang HW (2000) Action of antimicrobial peptides: two-state model. Biochemistry 39 (29):8347-8352

Hwang PM, Vogel HJ (1998) Structure-function relationships of antimicrobial peptides. Biochemistry and cell biology 76 (2-3):235-246

Imler J, Bulet P (2005) Antimicrobial peptides in Drosophila: structures, activities and gene regulation. In: Mechanisms of epithelial defense, vol 86. Karger Publishers, pp 1-21

Jenssen H (2005) Anti herpes simplex virus activity of lactoferrin/lactoferricin–an example of antiviral activity of antimicrobial protein/peptide. Cellular and Molecular Life Sciences CMLS 62 (24):3002-3013

Jenssen H, Andersen JH, Mantzilas D, Gutteberg TJ (2004) A wide range of medium-sized, highly cationic, α-helical peptides show antiviral activity against herpes simplex virus. Antiviral research 64 (2):119-126

Jenssen H, Hamill P, Hancock RE (2006) Peptide antimicrobial agents. Clinical microbiology reviews 19 (3):491-511

Jiravanichpaisal P, Lee SY, Kim Y-A, Andrén T, Söderhäll I (2007) Antibacterial peptides in hemocytes and hematopoietic tissue from freshwater crayfish Pacifastacus leniusculus: characterization and expression pattern. Developmental & Comparative Immunology 31 (5):441-455

Kayser O, Masihi KN, Kiderlen AF (2003) Natural products and synthetic compounds as immunomodulators. Expert review of anti-infective therapy 1 (2):319-335

Klotman ME, Chang TL (2006) Defensins in innate antiviral immunity. Nature Reviews Immunology 6 (6):447-456

REFERENCES

Konno K, Rangel M, Oliveira JS, dos Santos Cabrera MP, Fontana R, Hirata IY, Hide I, Nakata Y, Mori K, Kawano M (2007) Decoralin, a novel linear cationic α-helical peptide from the venom of the solitary eumenine wasp Oreumenes decoratus. Peptides 28 (12):2320-2327

Kragol G, Hoffmann R, Chattergoon MA, Lovas S, Cudic M, Bulet P, Condie BA, Rosengren KJ, Montaner LJ, Otvos L (2002) Identification of crucial residues for the antibacterial activity of the proline-rich peptide, pyrrhocoricin. European Journal of Biochemistry 269 (17):4226-4237

Kragol G, Lovas S, Varadi G, Condie BA, Hoffmann R, Otvos L (2001) The antibacterial peptide pyrrhocoricin inhibits the ATPase actions of DnaK and prevents chaperone-assisted protein folding. Biochemistry 40 (10):3016-3026

Kreil G Antimicrobial peptides from amphibian skin: an overview. In: Antimicrobial peptides. Ciba Foundation Symposium, 1994. vol 186. pp 77-90

Kuhn-Nentwig L, Müller J, Schaller J, Walz A, Dathe M, Nentwig W (2002) Cupiennin 1, a new family of highly basic antimicrobial peptides in the venom of the spider Cupiennius salei (Ctenidae). Journal of Biological Chemistry 277 (13):11208-11216

Laederach A, Andreotti AH, Fulton DB (2002) Solution and micelle-bound structures of tachyplesin I and its active aromatic linear derivatives. Biochemistry 41 (41):12359-12368

Lai R, Zheng Y-T, Shen J-H, Liu G-J, Liu H, Lee W-H, Tang S-Z, Zhang Y (2002) Antimicrobial peptides from skin secretions of Chinese red belly toad Bombina maxima. Peptides 23 (3):427-435

Lai Y, Gallo RL (2009) AMPed up immunity: how antimicrobial peptides have multiple roles in immune defense. Trends in immunology 30 (3):131-141

Lamberty M, Caille A, Landon C, Tassin-Moindrot S, Hetru C, Bulet P, Vovelle F (2001) Solution structures of the antifungal heliomicin and a selected variant with both antibacterial and antifungal activities. Biochemistry 40 (40):11995-12003

Lane J (2006) Mass vaccination and surveillance/containment in the eradication of smallpox. In: Mass Vaccination: Global Aspects—Progress and Obstacles. Springer, pp 17-29

Le Bihan T, Blochet J-É, Désormeaux A, Marion D, Pézolet M (1996) Determination of the secondary structure and conformation of puroindolines by infrared and Raman spectroscopy. Biochemistry 35 (39):12712-12722

REFERENCES

Lee DG, Kim HK, Am Kim S, Park Y, Park S-C, Jang S-H, Hahm K-S (2003) Fungicidal effect of indolicidin and its interaction with phospholipid membranes. Biochemical and biophysical research communications 305 (2):305-310

Lee DG, Park Y, Jin I, Hahm KS, Lee HH, Moon YH, Woo ER (2004) Structure–antiviral activity relationships of cecropin A-magainin 2 hybrid peptide and its analogues. Journal of Peptide Science 10 (5):298-303

Lehmann J, Retz M, Sidhu SS, Suttmann H, Sell M, Paulsen F, Harder J, Unteregger G, Stöckle M (2006) Antitumor activity of the antimicrobial peptide magainin II against bladder cancer cell lines. European urology 50 (1):141-147

Lehrer RI, Ganz T (2002) Cathelicidins: a family of endogenous antimicrobial peptides. Current opinion in hematology 9 (1):18-22

Leite JRS, Silva LP, Rodrigues MIS, Prates MV, Brand GD, Lacava BM, Azevedo RB, Bocca AL, Albuquerque S, Bloch C (2005) Phylloseptins: a novel class of anti-bacterial and anti-protozoan peptides from the Phyllomedusa genus. Peptides 26 (4):565-573

Levashina EA, Ohresser S, Bulet P, Reichhart JM, Hetru C, Hoffmann JA (1995) Metchnikowin, a novel immune-inducible proline-rich peptide from Drosophila with antibacterial and antifungal properties. European Journal of Biochemistry 233 (2):694-700

Li QF, 李祈福, Ou-Yang GL, 欧阳高亮, Li CY, Hong SG, 洪水根 (2000) Effects of tachyplesin on the morphology and ultrastructure of human gastric carcinoma cell line BGC-823.

Li W-F, Ma G-X, Zhou X-X (2006) Apidaecin-type peptides: biodiversity, structure–function relationships and mode of action. Peptides 27 (9):2350-2359

López-García B, Pérez-Payá E, Marcos JF (2002) Identification of novel hexapeptides bioactive against phytopathogenic fungi through screening of a synthetic peptide combinatorial library. Applied and environmental microbiology 68 (5):2453-2460

Lorin C, Saidi H, Belaid A, Zairi A, Baleux F, Hocini H, Bélec L, Hani K, Tangy F (2005) The antimicrobial peptide dermaseptin S4 inhibits HIV-1 infectivity in vitro. Virology 334 (2):264-275

Lynn MA, Kindrachuk J, Marr AK, Jenssen H, Panté N, Elliott MR, Napper S, Hancock RE, McMaster WR (2011) Effect of BMAP-28 antimicrobial peptides on Leishmania major promastigote and amastigote growth: role of leishmanolysin in parasite survival. PLoS Negl Trop Dis 5 (5):e1141

Mackintosh JA, Veal DA, Beattie AJ, Gooley AA (1998) Isolation from an

REFERENCES

ant Myrmecia gulosa of two inducible O-glycosylated proline-rich antibacterial peptides. Journal of Biological Chemistry 273 (11):6139-6143

Mai JC, Mi Z, Kim S-H, Ng B, Robbins PD (2001) A proapoptotic peptide for the treatment of solid tumors. Cancer Research 61 (21):7709-7712

Makovitzki A, Avrahami D, Shai Y (2006) Ultrashort antibacterial and antifungal lipopeptides. Proceedings of the National Academy of Sciences 103 (43):15997-16002

Mandard N, Bulet P, Caille A, Daffre S, Vovelle F (2002) The solution structure of gomesin, an antimicrobial cysteine-rich peptide from the spider. European Journal of Biochemistry 269 (4):1190-1198

Mangoni ML, Saugar JM, Dellisanti M, Barra D, Simmaco M, Rivas L (2005) Temporins, small antimicrobial peptides with leishmanicidal activity. Journal of Biological Chemistry 280 (2):984-990

Mannoor MS, Zhang S, Link AJ, McAlpine MC (2010) Electrical detection of pathogenic bacteria via immobilized antimicrobial peptides. Proceedings of the National Academy of Sciences 107 (45):19207-19212

Marcos JF, Muñoz A, Pérez-Payá E, Misra S, López-García B (2008) Identification and rational design of novel antimicrobial peptides for plant protection. Annu Rev Phytopathol 46:273-301

Matsuzaki K (1998) Magainins as paradigm for the mode of action of pore forming polypeptides. Biochimica et Biophysica Acta (BBA)-Reviews on Biomembranes 1376 (3):391-400

McManus AM, Dawson NF, Wade JD, Carrington LE, Winzor DJ, Craik DJ (2000) Three-dimensional structure of RK-1: a novel α-defensin peptide. Biochemistry 39 (51):15757-15764

Mcphee JB, Hancock RE (2005) Function and therapeutic potential of host defence peptides. Journal of Peptide Science 11 (11):677-687

Meng S, Xu H, Wang F (2010) Research advances of antimicrobial peptides and applications in food industry and agriculture. Current Protein and Peptide Science 11 (4):264-273

Meyer C, Reusser F (1967) A polypeptide antibacterial agent isolated fromTrichoderma viride. Experientia 23 (2):85-86

Miyata T, Tokunaga F, Yoneya T, Yoshikawa K, Iwanaga S, Niwa M, Takao T, Shimonishi Y (1989) Antimicrobial peptides, isolated from horseshoe crab hemocytes, tachyplesin II, and polyphemusins I and II: chemical structures and biological activity. Journal of biochemistry 106 (4):663-668

Montesinos E (2007) Antimicrobial peptides and plant disease control.

REFERENCES

FEMS microbiology letters 270 (1):1-11

Mookherjee N, Brown KL, Bowdish DM, Doria S, Falsafi R, Hokamp K, Roche FM, Mu R, Doho GH, Pistolic J (2006) Modulation of the TLR-mediated inflammatory response by the endogenous human host defense peptide LL-37. The Journal of Immunology 176 (4):2455-2464

Mor A, Van Huong N, Delfour A, Migliore-Samour D, Nicolas P (1991) Isolation, amino acid sequence and synthesis of dermaseptin, a novel antimicrobial peptide of amphibian skin. Biochemistry 30 (36):8824-8830

Moreno-Habel DA, Biglang-awa IM, Dulce A, Luu DD, Garcia P, Weers PM, Haas-Stapleton EJ (2012) Inactivation of the budded virus of Autographa californica M nucleopolyhedrovirus by gloverin. Journal of invertebrate pathology 110 (1):92-101

Murakami M, Ohtake T, Dorschner R, Gallo R (2002a) Cathelicidin antimicrobial peptides are expressed in salivary glands and saliva. Journal of dental research 81 (12):845-850

Murakami M, Ohtake T, Dorschner RA, Schittek B, Garbe C, Gallo RL (2002b) Cathelicidin anti-microbial peptide expression in sweat, an innate defense system for the skin. Journal of Investigative Dermatology 119 (5):1090-1095

Mushtaq S, Warner M, Cloke J, Afzal-Shah M, Livermore DM (2010) Performance of the Oxoid MIC Evaluator™ Strips compared with the Etest® assay and BSAC agar dilution. Journal of antimicrobial chemotherapy:dkq206

Mygind PH, Fischer RL, Schnorr KM, Hansen MT, Sönksen CP, Ludvigsen S, Raventós D, Buskov S, Christensen B, De Maria L (2005) Plectasin is a peptide antibiotic with therapeutic potential from a saprophytic fungus. Nature 437 (7061):975-980

Nakamura T, Furunaka H, Miyata T, Tokunaga F, Muta T, Iwanaga S, Niwa M, Takao T, Shimonishi Y (1988) Tachyplesin, a class of antimicrobial peptide from the hemocytes of the horseshoe crab (Tachypleus tridentatus). Isolation and chemical structure. Journal of Biological Chemistry 263 (32):16709-16713

Nguyen LT, Haney EF, Vogel HJ (2011) The expanding scope of antimicrobial peptide structures and their modes of action. Trends in biotechnology 29 (9):464-472

Nicolas P (2009) Multifunctional host defense peptides: intracellular-targeting antimicrobial peptides. Febs Journal 276 (22):6483-6496

Nijnik A, Madera L, Ma S, Waldbrook M, Elliott MR, Easton DM, Mayer

REFERENCES

ML, Mullaly SC, Kindrachuk J, Jenssen H (2010) Synthetic cationic peptide IDR-1002 provides protection against bacterial infections through chemokine induction and enhanced leukocyte recruitment. The Journal of Immunology 184 (5):2539-2550

Nizet V, Gallo RL (2003) Cathelicidins and innate defense against invasive bacterial infection. Scandinavian journal of infectious diseases 35 (9):670-676

Obeid M (2009) Anticancer activity of targeted proapoptotic peptides and chemotherapy is highly improved by targeted cell surface calreticulin–inducer peptides. Molecular cancer therapeutics 8 (9):2693-2707

Oh H-S, Ko S-S, Cho H, Lee K-H (2005) Design and synthesis of antibacterial pseudopeptides with a potent antibacterial activity and more improved stability from a short cationic antibacterial peptide. Bull Korean Chem Soc 26:161-164

Okumura K, Itoh A, Isogai E, Hirose K, Hosokawa Y, Abiko Y, Shibata T, Hirata M, Isogai H (2004) C-terminal domain of human CAP18 antimicrobial peptide induces apoptosis in oral squamous cell carcinoma SAS-H1 cells. Cancer letters 212 (2):185-194

Oppenheim F, Xu T, McMillian F, Levitz S, Diamond R, Offner G, Troxler R (1988) Histatins, a novel family of histidine-rich proteins in human parotid secretion. Isolation, characterization, primary structure, and fungistatic effects on Candida albicans. Journal of Biological Chemistry 263 (16):7472-7477

Oren Z, Shai Y (1998) Mode of action of linear amphipathic α-helical antimicrobial peptides. Peptide Science 47 (6):451-463

Orivel J, Redeker V, Le Caer J-P, Krier F, Revol-Junelles A-M, Longeon A, Chaffotte A, Dejean A, Rossier J (2001) Ponericins, new antibacterial and insecticidal peptides from the venom of the ant Pachycondyla goeldii. Journal of Biological Chemistry 276 (21):17823-17829

Otero-González AJ, Magalhães BS, Garcia-Villarino M, López-Abarrategui C, Sousa DA, Dias SC, Franco OL (2010) Antimicrobial peptides from marine invertebrates as a new frontier for microbial infection control. The FASEB Journal 24 (5):1320-1334

Otvos Jr L (2002) The short proline-rich antibacterial peptide family. Cellular and Molecular Life Sciences CMLS 59 (7):1138-1150

Otvos L, Bokonyi K, Varga I, Ertl HC, Hoffmann R, Bulet P, Otvos BI, Wade JD, Mcmanus AM, Craik DJ (2000) Insect peptides with improved protease-resistance protect mice against bacterial infection. Protein Science 9 (4):742-749

REFERENCES

Otvos L, Wade JD, Lin F, Condie BA, Hanrieder J, Hoffmann R (2005) Designer antibacterial peptides kill fluoroquinolone-resistant clinical isolates. Journal of medicinal chemistry 48 (16):5349-5359

Pan C-Y, Chen J-Y, Cheng Y-SE, Chen C-Y, Ni I-H, Sheen J-F, Pan Y-L, Kuo C-M (2007) Gene expression and localization of the epinecidin-1 antimicrobial peptide in the grouper (Epinephelus coioides), and its role in protecting fish against pathogenic infection. DNA and cell biology 26 (6):403-413

Pandey BK, Ahmad A, Asthana N, Azmi S, Srivastava RM, Srivastava S, Verma R, Vishwakarma AL, Ghosh JK (2010) Cell-selective lysis by novel analogues of melittin against human red blood cells and Escherichia coli. Biochemistry 49 (36):7920-7929

Papo N, Braunstein A, Eshhar Z, Shai Y (2004) Suppression of human prostate tumor growth in mice by a cytolytic d-, l-amino acid peptide membrane lysis, increased necrosis, and inhibition of prostate-specific antigen secretion. Cancer research 64 (16):5779-5786

Papo N, Seger D, Makovitzki A, Kalchenko V, Eshhar Z, Degani H, Shai Y (2006) Inhibition of tumor growth and elimination of multiple metastases in human prostate and breast xenografts by systemic inoculation of a host defense–like lytic peptide. Cancer research 66 (10):5371-5378

Papo N, Shahar M, Eisenbach L, Shai Y (2003) A novel lytic peptide composed of DL-amino acids selectively kills cancer cells in culture and in mice. Journal of Biological Chemistry 278 (23):21018-21023

Papo N, Shai Y (2003a) Can we predict biological activity of antimicrobial peptides from their interactions with model phospholipid membranes? Peptides 24 (11):1693-1703

Papo N, Shai Y (2003b) New lytic peptides based on the D, L-amphipathic helix motif preferentially kill tumor cells compared to normal cells. Biochemistry 42 (31):9346-9354

Park CB, Kim HS, Kim SC (1998) Mechanism of action of the antimicrobial peptide buforin II: buforin II kills microorganisms by penetrating the cell membrane and inhibiting cellular functions. Biochemical and biophysical research communications 244 (1):253-257

Park CH, Valore EV, Waring AJ, Ganz T (2001) Hepcidin, a urinary antimicrobial peptide synthesized in the liver. Journal of biological chemistry 276 (11):7806-7810

Park JM, Jung J-E, Lee BJ (1994) Antimicrobial peptides from the skin of a Korean frog, Rana rugosa. Biochemical and biophysical research

REFERENCES

communications 205 (1):948-954
Park T-J, Kim J-S, Choi S-S, Kim Y (2009) Cloning, expression, isotope labeling, purification, and characterization of bovine antimicrobial peptide, lactophoricin in Escherichia coli. Protein expression and purification 65 (1):23-29
Pazgier M, Li X, Lu W, Lubkowski J (2007) Human defensins: synthesis and structural properties. Current pharmaceutical design 13 (30):3096-3118
Pérez-Cordero JJ, Lozano JM, Cortés J, Delgado G (2011) Leishmanicidal activity of synthetic antimicrobial peptides in an infection model with human dendritic cells. Peptides 32 (4):683-690
Peters BM, Shirtliff ME, Jabra-Rizk MA (2010) Antimicrobial peptides: primeval molecules or future drugs? PLoS Pathog 6 (10):e1001067
Pillai A, Ueno S, Zhang H, Lee JM, Kato Y (2005) Cecropin P1 and novel nematode cecropins: a bacteria-inducible antimicrobial peptide family in the nematode Ascaris suum. Biochemical Journal 390 (1):207-214
Pinto MF, Fensterseifer IC, Migliolo L, Sousa DA, de Capdville G, Arboleda-Valencia JW, Colgrave ML, Craik DJ, Magalhães BS, Dias SC (2012) Identification and structural characterization of novel cyclotide with activity against an insect pest of sugar cane. Journal of Biological Chemistry 287 (1):134-147
Plan MRR, Saska I, Cagauan AG, Craik DJ (2008) Backbone cyclised peptides from plants show molluscicidal activity against the rice pest Pomacea canaliculata (golden apple snail). Journal of agricultural and food chemistry 56 (13):5237-5241
Ponti D, Mangoni ML, Mignogna G, Simmaco M, Barra D (2003) An amphibian antimicrobial peptide variant expressed in Nicotiana tabacum confers resistance to phytopathogens. Biochemical Journal 370 (1):121-127
Powers J-PS, Hancock RE (2003) The relationship between peptide structure and antibacterial activity. Peptides 24 (11):1681-1691
Powers J-PS, Rozek A, Hancock RE (2004) Structure–activity relationships for the β-hairpin cationic antimicrobial peptide polyphemusin I. Biochimica et Biophysica Acta (BBA)-Proteins and Proteomics 1698 (2):239-250
Powers J-PS, Tan A, Ramamoorthy A, Hancock RE (2005) Solution structure and interaction of the antimicrobial polyphemusins with lipid membranes. Biochemistry 44 (47):15504-15513
Pushpanathan M, Gunasekaran P, Rajendhran J (2013) Antimicrobial peptides: versatile biological properties. International journal of

REFERENCES

peptides 2013

Raj PA, Edgerton M (1995) Functional domain and poly-L-proline II conformation for candidacidal activity of bactenecin 5. FEBS letters 368 (3):526-530

Ramanathan B, Davis EG, Ross CR, Blecha F (2002) Cathelicidins: microbicidal activity, mechanisms of action, and roles in innate immunity. Microbes and Infection 4 (3):361-372

Rapaport D, Shai Y (1991) Interaction of fluorescently labeled pardaxin and its analogues with lipid bilayers. Journal of Biological Chemistry 266 (35):23769-23775

Reddy K, Aranha C, Gupta S, Yedery R (2004a) Evaluation of antimicrobial peptide nisin as a safe vaginal contraceptive agent in rabbits: in vitro and in vivo studies. Reproduction 128 (1):117-126

Reddy K, Yedery R, Aranha C (2004b) Antimicrobial peptides: premises and promises. International journal of antimicrobial agents 24 (6):536-547

Rees JA, Moniatte M, Bulet P (1997) Novel antibacterial peptides isolated from a European bumblebee, Bombus pascuorum (Hymenoptera, Apoidea). Insect biochemistry and molecular biology 27 (5):413-422

Rezansoff A, Hunter H, Jing W, Park I, Kim S, Vogel H (2005) Interactions of the antimicrobial peptide Ac-FRWWHR-NH2 with model membrane systems and bacterial cells. The Journal of peptide research 65 (5):491-501

Rifflet A, Gavalda S, Téné N, Orivel J, Leprince J, Guilhaudis L, Génin E, Vétillard A, Treilhou M (2012) Identification and characterization of a novel antimicrobial peptide from the venom of the ant Tetramorium bicarinatum. Peptides 38 (2):363-370

Rinaldi AC (2002) Antimicrobial peptides from amphibian skin: an expanding scenario: Commentary. Current opinion in chemical biology 6 (6):799-804

Risso A (2000) Leukocyte antimicrobial peptides: multifunctional effector molecules of innate immunity. Journal of leukocyte biology 68 (6):785-792

Risso A, Braidot E, Sordano MC, Vianello A, Macrì F, Skerlavaj B, Zanetti M, Gennaro R, Bernardi P (2002) BMAP-28, an antibiotic peptide of innate immunity, induces cell death through opening of the mitochondrial permeability transition pore. Molecular and cellular biology 22 (6):1926-1935

Risso A, Zanetti M, Gennaro R (1998) Cytotoxicity and apoptosis mediated by two peptides of innate immunity. Cellular immunology 189

REFERENCES

(2):107-115

Rizzo V, Stankowski S, Schwarz G (1987) Alamethicin incorporation in lipid bilayers: a thermodynamic study. Biochemistry 26 (10):2751-2759

Robinson WE, McDougall B, Tran D, Selsted ME (1998) Anti-HIV-1 activity of indolicidin, an antimicrobial peptide from neutrophils. Journal of leukocyte biology 63 (1):94-100

Rogers L (1928) The inhibiting effect of Streptococcus lactis on Lactobacillus bulgaricus. Journal of bacteriology 16 (5):321

Romeo D, Skerlavaj B, Bolognesi M, Gennaro R (1988) Structure and bactericidal activity of an antibiotic dodecapeptide purified from bovine neutrophils. Journal of Biological Chemistry 263 (20):9573-9575

Ross DJ, Cole AM, Yoshioka D, Park AK, Belperio JA, Laks H, Strieter RM, Lynch III JP, Kubak B, Ardehali A (2004) Increased bronchoalveolar lavage human β-defensin type 2 in bronchiolitis obliterans syndrome after lung transplantation. Transplantation 78 (8):1222-1224

Rozek A, Friedrich CL, Hancock RE (2000a) Structure of the bovine antimicrobial peptide indolicidin bound to dodecylphosphocholine and sodium dodecyl sulfate micelles. Biochemistry 39 (51):15765-15774

Rozek T, Waugh RJ, Steinborner ST, Bowie JH, Tyler MJ, Wallace JC (1998) The maculatin peptides from the skin glands of the tree frog Litoria genimaculata: a comparison of the structures and antibacterial activities of maculatin 1.1 and caerin 1.1. Journal of Peptide Science 4 (2):111-115

Rozek T, Wegener KL, Bowie JH, Olver IN, Carver JA, Wallace JC, Tyler MJ (2000b) The antibiotic and anticancer active aurein peptides from the Australian Bell Frogs Litoria aurea and Litoria raniformis. European Journal of Biochemistry 267 (17):5330-5341

Sai KP, Jagannadham MV, Vairamani M, Raju NP, Devi AS, Nagaraj R, Sitaram N (2001) Tigerinins: novel antimicrobial peptides from the Indian frogRana tigerina. Journal of Biological Chemistry 276 (4):2701-2707

Salomon R, Farías RN (1992) Microcin 25, a novel antimicrobial peptide produced by Escherichia coli. Journal of bacteriology 174 (22):7428-7435

Salzet M (2002) Antimicrobial peptides are signaling molecules. Trends in immunology 23 (6):283-284

Saravanan R, Li X, Lim K, Mohanram H, Peng L, Mishra B, Basu A, Lee

REFERENCES

JM, Bhattacharjya S, Leong SSJ (2014) Design of short membrane selective antimicrobial peptides containing tryptophan and arginine residues for improved activity, salt-resistance, and biocompatibility. Biotechnology and bioengineering 111 (1):37-49

Schibli DJ, Hwang PM, Vogel HJ (1999) Structure of the antimicrobial peptide tritrpticin bound to micelles: a distinct membrane-bound peptide fold. Biochemistry 38 (51):16749-16755

Schneider T, Kruse T, Wimmer R, Wiedemann I, Sass V, Pag U, Jansen A, Nielsen AK, Mygind PH, Raventós DS (2010) Plectasin, a fungal defensin, targets the bacterial cell wall precursor Lipid II. Science 328 (5982):1168-1172

Scocchi M, Tossi A, Gennaro R (2011) Proline-rich antimicrobial peptides: converging to a non-lytic mechanism of action. Cellular and Molecular Life Sciences 68 (13):2317-2330

Scocchi M, Wang S, Zanetti M (1997) Structural organization of the bovine cathelicidin gene family and identification of a novel member. FEBS letters 417 (3):311-315

Scott MG, Davidson DJ, Gold MR, Bowdish D, Hancock RE (2002) The human antimicrobial peptide LL-37 is a multifunctional modulator of innate immune responses. The Journal of Immunology 169 (7):3883-3891

Scott MG, Hancock RE (2000) Cationic antimicrobial peptides and their multifunctional role in the immune system. Critical Reviews™ in Immunology 20 (5)

Selsted ME, Miller SI, Henschen AH, Ouellette AJ (1992) Enteric defensins: antibiotic peptide components of intestinal host defense. The Journal of cell biology 118 (4):929-936

Selsted ME, Ouellette AJ (2005) Mammalian defensins in the antimicrobial immune response. Nature immunology 6 (6):551-557

Shai Y (1999) Mechanism of the binding, insertion and destabilization of phospholipid bilayer membranes by α-helical antimicrobial and cell non-selective membrane-lytic peptides. Biochimica et Biophysica Acta (BBA)-Biomembranes 1462 (1):55-70

Shai Y (2002) Mode of action of membrane active antimicrobial peptides. Peptide Science 66 (4):236-248

Shai Y, Bach D, Yanovsky A (1990) Channel formation properties of synthetic pardaxin and analogues. Journal of Biological Chemistry 265 (33):20202-20209

Shai Y, Oren Z (2001) From "carpet" mechanism to de-novo designed diastereomeric cell-selective antimicrobial peptides. Peptides 22 (10):1629-1641

REFERENCES

Silphaduang U, Noga E Peptide antibiotics in mast cells of fish. In: FASEB JOURNAL, 2002. vol 5. FEDERATION AMER SOC EXP BIOL 9650 ROCKVILLE PIKE, BETHESDA, MD 20814-3998 USA, pp A1225-A1225

Silva PI, Daffre S, Bulet P (2000) Isolation and characterization of gomesin, an 18-residue cysteine-rich defense peptide from the spider Acanthoscurria gomesiana hemocytes with sequence similarities to horseshoe crab antimicrobial peptides of the tachyplesin family. Journal of Biological Chemistry 275 (43):33464-33470

Simmaco M, Mignogna G, Barra D (1998) Antimicrobial peptides from amphibian skin: what do they tell us? Peptide Science 47 (6):435-450

Simmaco M, Mignogna G, Barra D, Bossa F (1993) Novel antimicrobial peptides from skin secretion of the European frog Rana esculenta. FEBS letters 324 (2):159-161

Simmaco M, Mignogna G, Barra D, Bossa F (1994) Antimicrobial peptides from skin secretions of Rana esculenta. Molecular cloning of cDNAs encoding esculentin and brevinins and isolation of new active peptides. Journal of Biological Chemistry 269 (16):11956-11961

Simmaco M, Mignogna G, Canofeni S, Miele R, Mangoni ML, Barra D (1996) Temporins, antimicrobial peptides from the European red frog Rana temporaria. European Journal of Biochemistry 242 (3):788-792

Sitaram N, Nagaraj R (1999) Interaction of antimicrobial peptides with biological and model membranes: structural and charge requirements for activity. Biochimica et Biophysica Acta (BBA)-Biomembranes 1462 (1):29-54

Skerlavaj B, Gennaro R, Bagella L, Merluzzi L, Risso A, Zanetti M (1996) Biological characterization of two novel cathelicidin-derived peptides and identification of structural requirements for their antimicrobial and cell lytic activities. Journal of Biological Chemistry 271 (45):28375-28381

Soares JW, Mello CM Antimicrobial peptides: a review of how peptide structure impacts antimicrobial activity. In: Optical Technologies for Industrial, Environmental, and Biological Sensing, 2004. International Society for Optics and Photonics, pp 20-27

Stciner H, Hultmark D, Engstrom A, Bennich H, Barman H (1981) Séquence and spe-cificity of two antibacterial proteins involved in insect immunity. Nature 292:246-248

Steinberg DA, Hurst MA, Fujii CA, Kung A, Ho J, Cheng F, Loury DJ,

REFERENCES

Fiddes JC (1997) Protegrin-1: a broad-spectrum, rapidly microbicidal peptide with in vivo activity. Antimicrobial agents and chemotherapy 41 (8):1738-1742

Strøm MB, Haug BE, Rekdal Ø, Skar ML, Stensen W, Svendsen JS (2002) Important structural features of 15-residue lactoferricin derivatives and methods for improvement of antimicrobial activity. Biochemistry and Cell Biology 80 (1):65-74

Strominger JL (2009) Animal antimicrobial peptides: ancient players in innate immunity. The Journal of Immunology 182 (11):6633-6634

Subbalakshmi C, Sitaram N (1998) Mechanism of antimicrobial action of indolicidin. FEMS microbiology letters 160 (1):91-96

Sutyak KE, Anderson RA, Dover SE, Feathergill KA, Aroutcheva AA, Faro S, Chikindas ML (2008) Spermicidal activity of the safe natural antimicrobial peptide subtilosin. Infectious diseases in obstetrics and gynecology 2008

Tan XW, Lakshminarayanan R, Liu SP, Goh E, Tan D, Beuerman RW, Mehta JS (2012) Dual functionalization of titanium with vascular endothelial growth factor and β-defensin analog for potential application in keratoprosthesis. Journal of Biomedical Materials Research Part B: Applied Biomaterials 100 (8):2090-2100

Tanaka S, Nakamura T, Morita T, Iwanaga S (1982) Limulus anti-LPS factor: An anticoagulant which inhibits the endotoxin-mediated activation of Limulus coagulation system. Biochemical and biophysical research communications 105 (2):717-723

Thompson SA, Tachibana K, Nakanishi K, Kubota I (1986) Melittin-like peptides from the shark-repelling defense secretion of the sole Pardachirus pavoninus. Science 233 (4761):341-343

Thouzeau C, Le Maho Y, Froget G, Sabatier L, Le Bohec C, Hoffmann JA, Bulet P (2003) Spheniscins, avian β-defensins in preserved stomach contents of the king penguin, Aptenodytes patagonicus. Journal of Biological Chemistry 278 (51):51053-51058

Tossi A, Sandri L, Giangaspero A (2000) Amphipathic, α-helical antimicrobial peptides. Peptide Science 55 (1):4-30

Tossi A, Scocchi M, Zanetti M, Storici P, Gennaro R (1995) PMAP-37, a Novel Antibacterial Peptide from Pig Myeloid Cells. European Journal of Biochemistry 228 (3):941-946

Travis SM, Anderson NN, Forsyth WR, Espiritu C, Conway BD, Greenberg EP, McCray PB, Lehrer RI, Welsh MJ, Tack BF (2000) Bactericidal activity of mammalian cathelicidin-derived peptides. Infection and immunity 68 (5):2748-2755

Vallespi MG, Glaria LA, Reyes O, Garay HE, Ferrero J, Araña MJ (2000) A

REFERENCES

Limulus antilipopolysaccharide factor-derived peptide exhibits a new immunological activity with potential applicability in infectious diseases. Clinical and diagnostic laboratory immunology 7 (4):669-675

van der Does AM, Bergman P, Agerberth B, Lindbom L (2012) Induction of the human cathelicidin LL-37 as a novel treatment against bacterial infections. Journal of leukocyte biology 92 (4):735-742. doi:10.1189/jlb.0412178

van der Weerden NL, Bleackley MR, Anderson MA (2013) Properties and mechanisms of action of naturally occurring antifungal peptides. Cellular and molecular life sciences 70 (19):3545-3570

VanCompernolle SE, Taylor RJ, Oswald-Richter K, Jiang J, Youree BE, Bowie JH, Tyler MJ, Conlon JM, Wade D, Aiken C (2005) Antimicrobial peptides from amphibian skin potently inhibit human immunodeficiency virus infection and transfer of virus from dendritic cells to T cells. Journal of virology 79 (18):11598-11606

Vernon LP, Evett GE, Zeikus RD, Gray WR (1985) A toxic thionin from Pyrularia pubera: purification, properties, and amino acid sequence. Archives of biochemistry and biophysics 238 (1):18-29

Vizioli J, Bulet P, Hoffmann JA, Kafatos FC, Müller H-M, Dimopoulos G (2001) Gambicin: a novel immune responsive antimicrobial peptide from the malaria vector Anopheles gambiae. Proceedings of the National Academy of Sciences 98 (22):12630-12635

Vizioli J, Salzet M (2002) Antimicrobial peptides versus parasitic infections? Trends in parasitology 18 (11):475-476

Vouldoukis I, Shai Y, Nicolas P, Mor A (1996) Broad spectrum antibiotic activity of skin-PYY. FEBS letters 380 (3):237-240

Wachinger M, Kleinschmidt A, Winder D, von Pechmann N, Ludvigsen A, Neumann M, Holle R, Salmons B, Erfle V, Brack-Werner R (1998) Antimicrobial peptides melittin and cecropin inhibit replication of human immunodeficiency virus 1 by suppressing viral gene expression. Journal of General Virology 79 (4):731-740

Wang CK, Colgrave ML, Gustafson KR, Ireland DC, Goransson U, Craik DJ (2007) Anti-HIV cyclotides from the Chinese medicinal herb Viola yedoensis. Journal of natural products 71 (1):47-52

Wang G, Mishra B, Lau K, Lushnikova T, Golla R, Wang X (2015) Antimicrobial peptides in 2014. Pharmaceuticals 8 (1):123-150

Wang G, Watson KM, Buckheit RW (2008) Anti-human immunodeficiency virus type 1 activities of antimicrobial peptides derived from human and bovine cathelicidins. Antimicrobial agents and

REFERENCES

chemotherapy 52 (9):3438-3440

Wang G, Watson KM, Peterkofsky A, Buckheit RW (2010a) Identification of novel human immunodeficiency virus type 1-inhibitory peptides based on the antimicrobial peptide database. Antimicrobial agents and chemotherapy 54 (3):1343-1346

Wang Y-D, Kung C-W, Chen J-Y (2010b) Antiviral activity by fish antimicrobial peptides of epinecidin-1 and hepcidin 1–5 against nervous necrosis virus in medaka. Peptides 31 (6):1026-1033

Wei L, Yang J, He X, Mo G, Hong J, Yan X, Lin D, Lai R (2013) Structure and function of a potent lipopolysaccharide-binding antimicrobial and anti-inflammatory peptide. Journal of medicinal chemistry 56 (9):3546-3556

Wei S-Y, Wu J-M, Kuo Y-Y, Chen H-L, Yip B-S, Tzeng S-R, Cheng J-W (2006) Solution structure of a novel tryptophan-rich peptide with bidirectional antimicrobial activity. Journal of bacteriology 188 (1):328-334

Weir GM, Liwski RS, Mansour M (2011) Immune modulation by chemotherapy or immunotherapy to enhance cancer vaccines. Cancers 3 (3):3114-3142

Wiegand I, Hilpert K, Hancock RE (2008) Agar and broth dilution methods to determine the minimal inhibitory concentration (MIC) of antimicrobial substances. Nature protocols 3 (2):163-175

Wong JH, Ng TB (2005) Sesquin, a potent defensin-like antimicrobial peptide from ground beans with inhibitory activities toward tumor cells and HIV-1 reverse transcriptase. Peptides 26 (7):1120-1126

Wu M, Hancock RE (1999) Interaction of the cyclic antimicrobial cationic peptide bactenecin with the outer and cytoplasmic membrane. Journal of Biological Chemistry 274 (1):29-35

Xiao Y, Cai Y, Bommineni YR, Fernando SC, Prakash O, Gilliland SE, Zhang G (2006) Identification and functional characterization of three chicken cathelicidins with potent antimicrobial activity. Journal of Biological Chemistry 281 (5):2858-2867

Xu J, Shi L, Zhou X, Xiao Z (2003) [Contraceptive efficacy of bioadhesive nonoxynol-9 gel: comparison with nonoxynol-9 suppository]. Zhonghua Fu Chan Ke Za Zhi 38 (10):629-631

Yan L, Adams ME (1998) Lycotoxins, antimicrobial peptides from venom of the wolf spiderLycosa carolinensis. Journal of Biological Chemistry 273 (4):2059-2066

Yang Q-Z, Wang C, Lang L, Zhou Y, Wang H, Shang D-J (2013) Design of potent, non-toxic anticancer peptides based on the structure of the antimicrobial peptide, temporin-1CEa. Archives of pharmacal

REFERENCES

research 36 (11):1302-1310

Yau W-M, Wimley WC, Gawrisch K, White SH (1998) The preference of tryptophan for membrane interfaces. Biochemistry 37 (42):14713-14718

Yeaman MR, Yount NY (2003) Mechanisms of Antimicrobial Peptide Action and Resistance. Pharmacological Reviews 55 (1):27-55. doi:10.1124/pr.55.1.2

Yeung AT, Gellatly SL, Hancock RE (2011) Multifunctional cationic host defence peptides and their clinical applications. Cellular and Molecular Life Sciences 68 (13):2161-2176

Zairi A, Belaïd A, Gahbiche A, Hani K (2005) Spermicidal activity of dermaseptins. Contraception 72 (6):447-453

Zampa MF, Araújo IMdS, dos Santos Júnior JR, Zucolotto V, Leite JRdS, Eiras C (2012) Development of a novel biosensor using cationic antimicrobial peptide and nickel phthalocyanine ultrathin films for electrochemical detection of dopamine. International journal of analytical chemistry 2012

Zanetti M (2004) Cathelicidins, multifunctional peptides of the innate immunity. Journal of leukocyte biology 75 (1):39-48

Zanetti M, Gennaro R, Romeo D (1995) Cathelicidins: a novel protein family with a common proregion and a variable C-terminal antimicrobial domain. FEBS letters 374 (1):1-5

Zasloff M (1987) Magainins, a class of antimicrobial peptides from Xenopus skin: isolation, characterization of two active forms, and partial cDNA sequence of a precursor. Proceedings of the National Academy of Sciences 84 (15):5449-5453

Zasloff M (2002) Antimicrobial peptides of multicellular organisms. nature 415 (6870):389-395

Zasloff M (2006) Inducing endogenous antimicrobial peptides to battle infections. Proceedings of the National Academy of Sciences 103 (24):8913-8914

Zhang L, Scott MG, Yan H, Mayer LD, Hancock RE (2000) Interaction of polyphemusin I and structural analogs with bacterial membranes, lipopolysaccharide, and lipid monolayers. Biochemistry 39 (47):14504-14514

Zwaal R, Comfurius P, Bevers E (2005) Surface exposure of phosphatidylserine in pathological cells. Cellular and Molecular Life Sciences CMLS 62 (9):971-988

Zwaal RF, Schroit AJ (1997) Pathophysiologic implications of membrane phospholipid asymmetry in blood cells. Blood 89 (4):1121-1132

ABOUT THE AUTHOR

Dr. Aqeel Ahmad

Dr. Aqeel Ahmad is currently working as a lecturer in College of Medicine, Shaqra University, KSA. He did M.Sc. in Biochemistry at Jamia Hamdard, New Delhi, India and obtained a Ph.D. in Biochemistry from CSIR-Central Drug Research Institute, Lucknow India/CSJM University Kanpur, India. He had worked as a Postdoctoral Researcher at Faculty of Medicine, University of Helsinki, Finland and at Department of Biomedical Engineering and Computational Science, Aalto University, Finland. He has been awarded Junior Research Fellowship by CSIR, India and DST-young Scientist fellowship by Department of Science and Technology, Government of India and received UP-CST Young Scientist Award in 2011 by Uttar Pradesh Council of Science and Technology, India. He has published several scientific papers in reputed international journals, Book chapters and got a patent in protein/peptide field.

Dr. Abdulrahman M Alshahrani

Dr. Abdulrahman M Alshahrani is currently working as a Dean at College of Medicine, Shaqra University, Saudi Arabia. He is also working as a consultant in neurology since 2002. He obtained his basic degree in Medicine (MBBS) from college of Medicine, King Saud University, Riyadh, Saudi Arabia. He received fellowship in the field of neurology in 1999. He has published several scientific papers in various international journals.

www.ingramcontent.com/pod-product-compliance
Lightning Source LLC
Chambersburg PA
CBHW061150180526
45170CB00002B/708